Galina Filipuk, Andrzej Kozłowski
Analysis with Mathematica®

Also of Interest

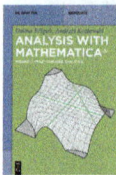

Analysis with Mathematica®. Volume 2: Multi-variable Calculus
Galina Filipuk, Andrzej Kozłowski, 2020
ISBN 978-3-11-066038-8, e-ISBN (PDF) 978-3-11-066039-5,
e-ISBN (EPUB) 978-3-11-066041-8

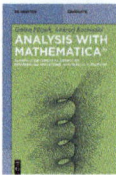

Analysis with Mathematica®. Volume 3: Differential Geometry, Differential Equations, and Special Functions
Galina Filipuk, Andrzej Kozłowski, 2022
ISBN 978-3-11-077454-2, e-ISBN (PDF) 978-3-11-077464-1,
e-ISBN (EPUB) 978-3-11-077475-7

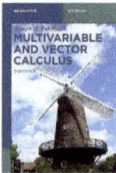

Multivariable and Vector Calculus
Joseph D. Fehribach, 2024
ISBN 978-3-11-139238-7, e-ISBN (PDF) 978-3-11-139348-3,
e-ISBN (EPUB) 978-3-11-139428-2

Scientific Computing. For Scientists and Engineers
Timo Heister, Leo Rebholz, 2023
ISBN 978-3-11-099961-7, e-ISBN (PDF) 978-3-11-098845-1,
e-ISBN (EPUB) 978-3-11-098875-8

Real Analysis – An Introduction.
Mathematical Arguments and Elementary Proof Techniques
Michael Cullinane, 2025
ISBN 978-3-11-142928-1, e-ISBN (PDF) 978-3-11-142956-4,
e-ISBN (EPUB) 978-3-11-142994-6

Galina Filipuk, Andrzej Kozłowski

Analysis with Mathematica®

—

Volume 1: Single Variable Calculus

2nd edition

DE GRUYTER

Mathematics Subject Classification 2020
97I20, 97I30, 97I40, 97I50, 97N80

Authors

Dr. hab. Galina Filipuk
University of Warsaw
Faculty of Mathematics, Informatics and Mechanics
Banacha 2
02-097 Warsaw
Poland
filipuk@mimuw.edu.pl

Dr. Andrzej Kozłowski
University of Warsaw
Faculty of Mathematics, Informatics and Mechanics
Banacha 2
02-097 Warsaw
Poland
akoz@mimuw.edu.pl

ISBN 978-3-11-152214-2
e-ISBN (PDF) 978-3-11-153306-3
e-ISBN (EPUB) 978-3-11-153315-5

Library of Congress Control Number: 2025941722

Bibliographic information published by the Deutsche Nationalbibliothek
The Deutsche Nationalbibliothek lists this publication in the Deutsche Nationalbibliografie; detailed bibliographic data are available on the Internet at http://dnb.dnb.de.

© 2026 Walter de Gruyter GmbH, Berlin/Boston, Genthiner Straße 13, 10785 Berlin
Cover image: PhonlamaiPhoto / iStock / Getty Images Plus
Typesetting: VTeX UAB, Lithuania

www.degruyter.com
Questions about General Product Safety Regulation:
productsafety@degruyterbrill.com

Contents

Preface —— IX

1	**Number systems —— 1**	
1.1	Sets —— 1	
1.2	Domains —— 2	
1.3	Assumptions in Mathematica® —— 7	
1.4	Quantifiers —— 10	
1.5	Complex numbers —— 12	
1.6	Real numbers —— 13	
1.7	Infinities —— 16	
1.8	Numbers in different bases —— 18	
1.9	Interval arithmetic —— 18	
1.10	Telescoping in Mathematica® —— 19	
1.11	Integers and the Principle of Mathematical Induction —— 20	
1.11.1	Example —— 21	
1.11.2	Example —— 22	
1.12	Algebraic equations and algebraic numbers —— 30	
1.12.1	Example —— 34	
1.13	Non-algebraic equations —— 36	
1.14	Sequences of real numbers and their limits —— 37	
1.14.1	Example —— 42	
1.14.2	Example: the number e —— 42	
1.14.3	Example —— 43	
1.14.4	Example: the Euler–Mascheroni constant —— 44	
1.15	Supremum and infimum —— 45	
1.15.1	Example —— 48	
1.15.2	Example —— 51	
1.15.3	Example —— 53	
2	**Recursive sequences, discrete dynamical systems and their limits —— 55**	
2.1	Example —— 59	
2.2	Example: the Fibonacci sequence —— 65	
2.3	Example —— 69	
2.4	Example: the secant method —— 70	
3	**Series —— 74**	
3.1	Sequences and series —— 74	
3.2	The functions Sum and NSum —— 76	
3.3	Absolute convergence —— 80	
3.4	Convergence of series with terms of constant signs —— 81	

3.4.1   Example —— 84
3.4.2   Example —— 86
3.4.3   Example —— 86
3.5   Convergence of series with terms of non-constant signs —— 87
3.5.1   Grouping of terms —— 87
3.5.2   Example —— 88
3.5.3   Example —— 89
3.5.4   Abel's summation formula —— 90
3.5.5   Dirichlet's and Abel's tests —— 91
3.5.6   Example —— 93
3.6   The function SumConvergence —— 95
3.6.1   Example —— 97
3.7   Riemann's theorem on conditionally convergent series —— 98
3.8   The Cauchy product of series —— 100
3.9   Divergent series —— 101
3.10   Power series —— 103

4   Limits of functions and continuity —— 109
4.1   Limits of functions —— 109
4.2   One-sided limits —— 110
4.3   Continuous functions —— 114
4.4   Discontinuous functions —— 117
4.4.1   Example: the Dirichlet function —— 120
4.5   The main theorems on continuous functions —— 122
4.5.1   Example —— 123
4.5.2   Example —— 124
4.5.3   Example —— 125
4.6   Inverse functions and their continuity —— 127
4.7   Example: recursive sequences and continuity —— 130
4.8   Uniform continuity and the Lipschitz property —— 131
4.8.1   Example —— 133

5   Differentiation —— 136
5.1   Difference quotient and derivative of a function —— 136
5.2   Differentiation in Mathematica® —— 144
5.2.1   Differentiation of expressions using D —— 145
5.2.2   Differentiation of functions using Derivative —— 147
5.2.3   Algebraic rules of differentiation —— 149
5.2.4   Example: user-defined derivative —— 150
5.3   Main properties of differentiable functions —— 151
5.3.1   Example: global and local extrema —— 153
5.3.2   Example —— 156

5.3.3	Example: the inverse function of the hyperbolic sine — **159**	
5.3.4	Example: number of roots of an equation — **161**	
5.3.5	Example: number of roots of an equation and the Lambert *W* function — **163**	
5.3.6	Example: implicit differentiation — **167**	
5.3.7	Example: the Schwarzian derivative — **168**	
5.3.8	Example: a tangent line to a curve — **168**	
5.4	Convex functions — **170**	
5.4.1	Jensen's inequality — **176**	

6	**Sequences and series of functions — 177**	
6.1	Power series continued — **177**	
6.1.1	Example — **178**	
6.1.2	Example — **179**	
6.2	Taylor polynomials and Taylor series — **181**	
6.2.1	Example — **187**	
6.2.2	Example — **188**	
6.2.3	Approximating functions by Taylor polynomials — **189**	
6.2.4	Example: rational approximation of $\sqrt{e}$ — **191**	
6.2.5	Example: illustration of approximation of functions with Taylor polynomials — **192**	
6.2.6	Example — **194**	
6.3	Convergence of sequences and series of functions — **200**	
6.3.1	Examples: pointwise, uniform and almost uniform convergence of function sequences — **202**	
6.3.2	Continuity and differentiability of limits and sums — **206**	
6.3.3	Examples: pointwise, uniform and almost uniform convergence of function series — **210**	
6.3.4	Example — **212**	

7	**Integration — 215**	
7.1	Indefinite integrals — **215**	
7.2	The Risch algorithm — **219**	
7.2.1	Differential algebras — **220**	
7.2.2	Example 1: integration of rational functions — **222**	
7.2.3	Example 2: the Risch algorithm for an exponential extension — **224**	
7.2.4	Limitations of Mathematica®'s integration — **225**	
7.3	The Riemann integral — **226**	
7.3.1	Using Integrate and NIntegrate with definite integrals — **233**	
7.3.2	Riemann sums — **235**	
7.4	Improper integrals — **237**	
7.4.1	Integrals over infinite intervals (improper integrals of the first type) — **238**	

7.4.2　　　Improper integrals of the first type and infinite sums —— **239**
7.4.3　　　Integrals of unbounded functions (improper integrals of the second type) —— **243**

8　　　LLMs with Mathematica —— 245
8.1　　　How to use LLMs from within Mathematica®? —— **248**
8.2　　　Example 1: Prime[n] vs 5n —— **252**
8.3　　　Example 2: a question of convergence —— **254**
8.4　　　When Mathematica® cannot solve a seemingly simple problem —— **255**
8.5　　　Asking about bugs —— **257**
8.6　　　More advanced example: the Risch algorithm —— **259**
8.7　　　Conclusion —— **263**

Bibliography —— 269

Index —— 271

Preface

Preface to the second edition

The new edition of the book contains numerous additions. New material has been included in every chapter, mostly in the form of new examples or additional explanations of mathematical concepts as well as the Wolfram Programming Language. We used Mathematica®'s version 14.2. Following the practice of Wolfram Research, we have adopted the name Wolfram Language for the programming language used by Mathematica®, while we continue to use Mathematica® to refer to the program itself. The distinction between these two may sometimes be blurred.

Brief descriptions of many Wolfram Language functions, which we believe may be useful to the readers of this text have also added. In some chapters we have also included more extensive discussions of certain aspects of the programming language that we found relevant (e. g., the discussion of the `Module` and `Block` constructions in Chapter 1). However, the most novel addition is the entirely new Chapter 8, devoted to a new feature of Mathematica® (and of Wolfram Language), which only appeared after the first edition was published. This new feature is the ability to use Artificial Intelligence, or, more accurately, Large Language Model chat bots from within Mathematica®. This is an exciting new development, which will play a big part in the future of mathematics education and, although it is still in its infancy, it is growing up remarkably fast and should be of interest to both students and educators.

Preface to the first edition

The main objective of this two-volume book is to convince the readers that Mathematica[1] can be used effectively as an aid in solving mathematical problems or at least to inspire the idea of a solution. Since we do not want to devote much space or time to repeating material covered in many other places, we will assume that the readers are already familiar with the basics of real analysis (e. g., as presented in the book [14] or in the more advanced book [12], which contains more than we need) and of Mathematica® (including some elements of the Wolfram Programming Language™ roughly of the same scope as in Sections 1–6 of [9]).

Our approach is to describe some of the theory and to guide the readers through solutions to a number of classroom problems. Most of these problems come from the lecture notes [13] for the course of analysis for computer science students, given at the University of Warsaw during 2011–2018. The problems we chose were the ones in which

---

[1] Wolfram Mathematica® is a system for modern technical computing, see http://www.wolfram.com/ for further information.

https://doi.org/10.1515/9783111533063-203

Mathematica® could be genuinely useful in a way that did not involve simply applying some standard algorithm or plotting a graph, but which required the students to think and come up with some mathematical idea. Sometimes this idea is a guess inspired by a Mathematica® calculation or a visual representation of the problem. In some special cases Mathematica® can only verify this guess and sometimes it can prove its correctness in full generality. In all such cases we normally still want to obtain a rigorous mathematical proof. In some cases we will provide such proofs in detail, although sometimes we will only sketch them or limit ourselves to a special case, leaving the details and greater generality to the readers.

We will sometimes also use Mathematica®'s abilities to solve mathematical problems which are too difficult for us to solve by hand (sometimes because of their complexity and sometimes because the solution requires mathematical knowledge that lies beyond the scope of this book). We believe that such examples are useful and interesting because they show how the strengths of Mathematica® (or other computer programs) can complement human weaknesses and vice versa. When using a computer program as an aid in mathematics it is important to understand what such programs are well suited to and what they are not, and also what kind of problems (including incorrect answers) one can run into. To learn to judiciously use a computer is becoming a necessity in many areas of science and mathematics and we hope our book will prove useful in this respect.

We shall frequently state well-known mathematical theorems and often omit their proofs. Usually we will refer the readers to books included in the bibliography (except when the proof is simple enough for the readers to supply it themselves). In some cases we will only give a "proof" based on Mathematica® or an illustration. There is a natural question that has been much discussed in recent years: when can results obtained by means of a computer software be regarded as mathematical proofs? The most popular view is that such results can be considered as proofs provided the source code of the mathematical software is open so that it can be examined to make sure that it is correct. In the case of commercial programs like Mathematica® the source code is not publicly available so one should consider results obtained by means of such programs as tentative and prove them by mathematical means or at least confirm their validity by means of other, independent programs. Since our purpose is to use Mathematica® as an educational tool, we shall not be concerned with the correctness of Mathematica®'s implementation of various algorithms and we will usually not try to describe them. However, the readers should be aware that the methods used by mathematical computer programs are usually very different from the ones used by humans. In addition, often the methods (algorithms) used by Mathematica® to obtain even quite elementary results depend on mathematical tools much more advanced than the scope of this book. Among examples of this kind are solving transcendental (non-polynomial) equations and inequalities, computing finite sums and sums of series, integration, etc.

We have been teaching analysis with Mathematica® at the University of Warsaw for computer science students for several years and we have found the above described approach quite effective. A typical mathematics course at the University of Warsaw consists of lectures at which the theory is taught and of accompanying problem classes. This book is primarily based on problem classes and therefore has a similar style. We only briefly sketch the necessary theoretical material and refer the readers for further details to standard textbooks, some of which can be found in the bibliography. Our book can also be seen as a supplement for any standard textbook on calculus or analysis and we shall cover standard first and second year calculus (analysis) topics, where we can benefit from using Mathematica® a lot. Volume 1 is devoted to single-variable calculus and volume 2 to the many-variable case. Our examples belong to pure mathematics and we do not consider any applications in physics and other sciences. Many textbooks exist which deal with such applications.

Although this book is based on a course of analysis for computer science students, it was actually a traditional mathematical course illustrated with examples from Mathematica® rather than a course of constructive or algorithmic mathematics (i. e., the kind of mathematics that Mathematica® itself makes use of). An excellent example of an introduction to analysis of the latter kind is [4], which we highly recommend. The difference between the constructive approach and the traditional approach is that the former avoids all indirect methods, such as "proofs by contradiction" which cannot be implemented on a computer. These are replaced, whenever possible, by "constructions", which in principle could be turned into computer programs. Not everything in analysis can be done constructively (and thus not everything can be implemented on a computer) and the extent to which this is possible is an active area of research. Although in this book we do not adopt a constructive approach, we only try to attract the attention of readers to such issues when they arise. We also do not consider the algorithms used by Mathematica® in detail (we recommend the already mentioned book [9] for some of such discussions, especially in connection with the Risch integration algorithm). We will make some comments on the algorithms from time to time in the hope that some readers will be inspired to pursue such topics further. Finally, there is another very important issue that we will only touch the surface of. It does not belong to traditional mathematics but is something of which one soon becomes aware of (often painfully) when one tries to apply Mathematica®'s symbolic capabilities to "real life" mathematical problems. This issue is computational complexity. In traditional mathematics a problem is often considered as solved as soon as a method (an algorithm) of solving it is found. These algorithms may, however, have such high complexity that in practice they allow only very simple cases to be solved in acceptable time. Some of the most impressive algorithms (e. g., quantifier elimination discussed in Chapter 1) have a very high complexity. This motivates the search for methods of solving problems quite different from the standard ones (an example of this is the standard algorithm for integrating rational functions, taught in all traditional Calculus courses, which is unusable in practice due to its need to factorize arbitrary polynomials). Again, in this book we only point out the existence of these important issues.

As mentioned earlier, we assume some familiarity with the programming language used by Mathematica® (Wolfram Language™), which means its basic syntax, the structure of Mathematica® expressions, different forms of expressions (InputForm, TraditionalForm, FullForm), patterns and pattern matching, etc. Recent versions of Mathematica® can use the so called "free-form input", which allows the user to use "natural language" to tell Mathematica® to solve certain problems. Natural language has obvious advantages over formal languages. For example, Mathematica® has functions Element and MemberQ but only MemberQ[{1,2,3},1] and Element[1,Integers] will produce the answer True. If we use free-form input, the questions "is 1 an element of the set {1, 2, 3}?", "is 1 a member of the set {1, 2, 3}?" and "is 1 a member of the rationals?" will return the same answer True. Although free-form input is improving, it is still quite limited in scope and we have decided not to use it. Using a formal language has the advantage of clarity, which is essential when we want to do something more complicated than a single operation like solving an equation or plotting a graph. Of course there is a price. Not only do we have to use the correct syntax (for example, in the Wolfram Language different kind of brackets ( ), { }, [ ] have different meanings and must not be mixed up) but to use Mathematica® effectively one needs to use its built-in functions. However, there are many thousands of them, each having a large number of options. In practice it means that one of the most important Mathematica® skills is efficient use of its extensive documentation (the Wolfram Language & System Documentation Center™). Whenever the readers come across an unfamiliar function or an unfamiliar option of a familiar function they should use "?FunctionName" to obtain information about it. In fact, it is not necessary to remember the whole function name as long as Mathematica®'s autocomplete feature is turned on (it is by default). We expect the readers to look up the necessary information themselves but we will sometimes provide links to some selected help topics. We would also like to point out that this book is written in LaTeX and it is not possible to make the input and output exactly the way they would look in Mathematica®. Most of our input cells will resemble InputForm with a mixture of StandardForm, while the output cells will mostly resemble StandardForm. Large input and output cells are sometimes shown as graphics. We shall use the large letter I in both input and output to represent the complex number i, where $i^2 = -1$, although in StandardForm Mathematica® actually uses a special symbol for it. The same is true for the letter E for exp(1).

Although Mathematica® has become quite affordable recently, still not everybody has access to the latest version. All examples in this book should work in the same way with versions higher than 11.3 but this may not always be true for earlier ones. Some functions, such as DiscreteLimit, were introduced only in version 11.2 and will not work in earlier ones. Usually one can obtain the same results in earlier versions by means of a more complicated code. For example, DiscreteLimit can in many cases be replaced by Limit but not always (we give an example below). Sometimes (although comparatively rarely) the behavior of the built-in functions changes between different versions. For example, this happened to the function Limit: in Mathematica® 10 it always returns (by default) one-sided limits, in Mathematica® 11 it returns two-sided ones.

Thus, in Mathematica® 10, `Limit[1/n, n -> 0]` returns ∞, while in Mathematica® 11 it returns `Indeterminate`. This is because the left sided and the right sided limits are different: −∞ and ∞.

Apart from such differences, any version of Mathematica® later than 6.0 should be suitable for most of the Mathematica® code in this book. Earlier versions are much less suitable because the graphics, both static and interactive, will not work. We strongly encourage the readers to consult the documentation for more detailed and up-to-date information and possible further issues.

We also recommend reading this book from the beginning since many pieces of advice are scattered throughout the text (since the material can be grouped differently and sometimes we digress to some interesting related topics).

Finally, we would like to say a few words about our use of the word "function" throughout the book. We assume that all readers are familiar with the ordinary use of the word "function" in mathematics. Informally it is a "black box" mechanism that, for a given element of a set, usually a number x, produces an element of another set, e. g., x^2. More formally, a function or a mapping is a special case of a relation between two sets. The word function is also used in computer science (where it is a special case of what is called a "subroutine"). It is a program which, when supplied with certain input data, produces an output (the usual way to express this is by saying "it returns an output"). Applying a function to data (which must be given as function's arguments) and computing the result is called "evaluation". To be a function a subroutine should always return output. Subroutines that only change some data in computer's memory are not functions. Like mathematical functions, computer science functions have arguments and can be composed. The programming style based on writing programs as compositions of (preferably simple) functions is called "functional". This is the style of programming most suited to mathematics and Mathematica® programs written in this style (which implies avoiding looping constructions such as `Do`, `For`, `While`, which, since they do not produce output, are not functions) tend to be more efficient than those written in the more common "procedural" style. Throughout the book we use the word "function" in both mathematical and computer science sense without trying to distinguish them. We hope that after reading this introduction the readers will not find it confusing.

Although there exist a number of books on calculus with Mathematica®, we hope that ours offers a different perspective and will inspire the reader to experiment with other topics in analysis, which we have had to leave out. This book is primarily intended for educational purposes, but we believe that some parts of it can be of interest to researchers as well.

1 Number systems

Usually, courses of analysis begin with number systems. Real analysis begins with the real numbers, which are either introduced axiomatically [12, 1] or constructed from something more "basic", e. g., natural numbers or integers (which themselves are either introduced axiomatically [14] or defined in terms of sets [15]).

In this chapter we shall explain how Mathematica® deals with sets, numbers and sequences of numbers and consider many other related topics.

1.1 Sets

One of the first notions one has to learn to study modern mathematics is the notion of a set, since the natural language of modern mathematics is provided by set theory. Mathematica® uses the so called Wolfram Language, which does not have a built-in function corresponding to the mathematical notion of "set". It is important to remark that the built-in function "Set" means something completely different. A finite ordered set exists in Mathematica® as List. One can also study the properties of ordinary finite sets by using the fact that Mathematica® has a natural way to sort (or order) any list of objects. This is done either by the function Sort or by the function Union:

In[·]:= Sort[{b, 7, 3, "dog", "cat", a}]
Out[·]:= {3, 7, cat, dog, a, b}

In[·]:= Union[{1, 2, 1, 3, 6, 2, 2}]
Out[·]:= {1, 2, 3, 6}

In the example above the list contains objects of different kinds (symbols, numbers and strings of texts), but Mathematica® has a canonical way to arrange them. Because of this, any finite set of objects has a canonical representation as an ordered set so we can study sets by means of these canonical representations. If we want to check if two sets are equal, it is enough to check whether their canonical representations are equal. For example, the lists {1,2,3} and {2,1,3} are clearly not equal:

In[·]:= {1, 2, 3} == {2, 1, 3}
Out[·]:= False

but they represent the same sets, therefore applying Sort or Union will return True:

In[·]:= Union[{1, 2, 3}] == Union[{2, 1, 3}]
Out[·]:= True

Note that the function Join will not give the same result:

In[·]:= Join[{1, 2, 3}] == Join[{2, 1, 3}]
Out[·]:= False

https://doi.org/10.1515/9783111533063-001

since it just concatenates lists without sorting them:

In[·]:= Join[{2, 1, 3}, {1}]
Out[·]:= {2, 1, 3, 1}

Standard mathematical operations on finite sets (like union, intersection, complement) are implemented in Mathematica® as functions. For instance, Union applied to lists performs the usual sum operation on sets:

In[·]:= Union[{2, 3, 4}, {2, 3, 5}]
Out[·]:= {2, 3, 4, 5}

Intersection and Complement are other built in set-theoretic functions. They only work on lists.

Membership in a list is determined by the function MemberQ (which should not be confused with Element, see below):

In[·]:= MemberQ[{a, b, c, d, 1}, d]
Out[·]:= True

1.2 Domains

Certain sets of numbers are built in Mathematica® as "domains": Integers, Rationals, Reals, Complexes, Algebraics and Primes. As their names suggest, the domain Reals is used for real numbers, Complexes is for complex numbers, Algebraics is for numbers that solve polynomial equations with rational coefficients and so on. As expected, the usual inclusions of sets are preserved (that is, the domain of Integers is a subset of the domain of Rationals). There is one more domain, Booleans, which consists of symbols True and False. Many functions will give the answer True or False, for instance comparing two numbers and determining which one is greater or less. You can type in either 1 < 2 or Less[1,2] to get an output "True".

Domains are generally used together with the function Element, which allows one to test for membership in these domains (for checking whether a given number is an element of a certain domain) and often with certain functions, e. g., Simplify, FullSimplify, Reduce, Solve, Resolve, Minimize, Maximize, FindInstance. We will consider them later on.

Let us have a look at the following examples:

In[·]:= Element[2, Reals]
Out[·]:= True

In[·]:= Element[Sqrt[2], Rationals]
Out[·]:= False

As can be seen from these examples, the function Element takes two arguments, a number and a domain, and it is a Boolean function, which means that we can think of it as answering the following question: is the number an element of the domain? If the answer is not immediately obvious, Mathematica® returns the input unevaluated:

```
In[·]:= Element[(Sqrt[3] - Sqrt[2])*(Sqrt[2] + Sqrt[3]),
        Integers]
Out[·]:= (-√2 + √3) (√2 + √3) ∈ ℤ
```

This illustrates one of Mathematica®'s basic principles: when there is an input for which Mathematica® has no rules to apply, it returns the input. We can then apply to it additional transformations, not performed by default by the Mathematica® kernel. The simplest way is to use the function Simplify or its more powerful (but more time-consuming) variant FullSimplify:

```
In[·]:= Simplify[Element[(Sqrt[3] - Sqrt[2])*(Sqrt[2] +
        Sqrt[3]), Integers]]
Out[·]:= True
```

Instead of using Simplify or FullSimplify one can apply one of many built-in functions that Mathematica® has for transforming expressions. For example, in this case Expand will show us that the expression above is, in fact, 1, and then we can check whether it is an integer or not:

```
In[·]:= Expand[(Sqrt[3] - Sqrt[2])*(Sqrt[2] + Sqrt[3])]
Out[·]:= 1
```

```
In[·]:= Element[%, Integers]
Out[·]:= True
```

Some functions accept domains as one of their arguments. For example, to solve a given algebraic equation over the reals we can use

```
In[·]:= x /. Solve[x^3 == 1, x, Reals]
Out[·]:= {1}
```

In many cases if we do not specify the domain, Mathematica® assumes that it is working with complex numbers. For instance, just solving the cubic equation above without specifying the domain yields three roots:

```
In[·]:= x /. Solve[x^3 == 1, x]
Out[·]:= {1, -(-1)^{1/3}, (-1)^{2/3}}
```

We can also use a different method to obtain the real solutions: to find first the complex ones and then select only the real ones:

```
In[·]:= Select[x /. Solve[x^3 == 1, x], Element[#1, Reals] & ]
Out[·]:= {1}
```

Returning to the functions Simplify and FullSimplify, there are several useful options. We can add an option TimeConstraint to FullSimplify to control the time spent on evaluation. This forces Mathematica® to abort the evaluation once the time set in TimeConstraint is reached and to return the simplest form of the expression found so far.

What Simplify actually does is to try to simplify the given expression using built-in rules as well as possibly additional rules and information provided by the users. There is a default sense in which one expression is considered as simpler than another one.

The built-in notion of complexity of an expression (corresponding roughly to its LeafCount, which is based on the FullForm representation of expressions or the representation of expressions by means of the graph-theoretic notion of trees using the function TreeForm) will not always agree with what appears "simpler" to human eyes. For instance, Mathematica® by default considers the expression $\sqrt{xy}$ as "simpler" than $\sqrt{x}\,\sqrt{y}$:

In[·]:= {LeafCount[Sqrt[x*y]], LeafCount[Sqrt[x]*Sqrt[y]]}
Out[·]:= {7, 11}

In[·]:= FullForm[Sqrt[x*y]]
Out[·]//FullForm= Power[Times[x,y], Rational[1,2]]

In[·]:= FullForm[Sqrt[x]*Sqrt[y]]
Out[·]//FullForm= Times[Power[x, Rational[1, 2]], Power[y, Rational[1,2]]]

In fact, Mathematica®'s default notion of "simplicity" (or rather complexity) can be changed by the user by means of the options ComplexityFunction and TransformationFunctions. For example, Mathematica® cannot find any simpler form of the following expression by using the default ComplexityFunction:

In[·]:= FullSimplify[Sin[x]*Cos[x]]
Out[·]:= Cos[x] Sin[x]

However, suppose we want to find an expression without any cosines. For this purpose we need to define some function which will "penalize" the presence of cosines. For example, the complexity function below adds to the default ComplexityFunction (which is essentially LeafCount) three times the number of occurrences of cosines in the expression:

In[·]:= Simplify[Sin[x]*Cos[x], ComplexityFunction ->
          (LeafCount[#1] + 3*Count[#1, _Cos, Infinity] &)]
Out[·]:= $\frac{1}{2}$ Sin[2 x]

Of course, in this example instead of changing ComplexityFunction one can just use the built-in function TrigReduce:

In[·]:= TrigReduce[Sin[x]*Cos[x]]
Out[·]:= $\frac{1}{2}$ Sin[2 x]

Simplify has another important option TransformationFunctions. By default, Simplify uses a number of transformations, which replace an expression by an equivalent expression. However, some useful functions are not used, since they are generally time-consuming. For example, the function Factor, which tries to factorize polynomials,

In[·]:= Factor[x^2 - 4]
Out[·]:= (-2 + x)(2 + x)

In[·]:= Factor[x^2 - 2]
Out[·]:= -2 + x²

In[·]:= Factor[x^2 - 2, Extension -> Sqrt[2]]
Out[·]:= -((√-2 - x)(√2 + x))

has a very high time complexity, and therefore this function is not used by default. However, one can always add it to all functions used automatically as follows:

In[·]:= FullSimplify[(x^4 - 16)/(2 + x),
        TransformationFunctions -> {Automatic, Factor}]
Out[·]:= (-2 + x)(4 + x²)

Compare this with FullSimplify with default options:

In[·]:= FullSimplify[(x^4 - 16)/(2 + x)]
Out[·]:= $\dfrac{-16 + x^4}{2 + x}$

Mathematica® can also expand certain expressions by using the function FunctionExpand:

In[·]:= FunctionExpand[Sin[2*ArcTan[x]]]
Out[·]:= $\dfrac{2x}{1 + x^2}$

Mathematica® considers the following expression as simpler:

In[·]:= FullSimplify[Sin[2*ArcTan[x]]]
Out[·]:= Sin[2 ArcTan[x]]

In[·]:= LeafCount[Sin[2*ArcTan[x]]]
Out[·]:= 5

In[·]:= TreeForm[Sin[2*ArcTan[x]]]

Out[·]//TreeForm=

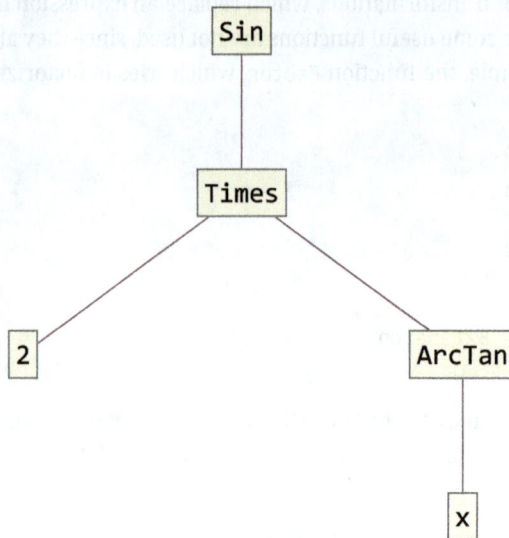

Figure 1.1

Compare with

In[·]:= `LeafCount[2*(x/(1 + x^2))]`
Out[·]:= 10

If we want `FullSimplify` to convert our expression to one that does not involve the `Sin` function, the simplest way is to define a function that counts the number of times `Sin` occurs in an expression (at all levels), and use it as a measure of complexity.

In[·]:= `f[x_] := Count[x, Sin, Infinity, Heads -> True]`

In[·]:= `f[Sin[x]]`
Out[·]:= 1

In[·]:= `f[Sin[2*ArcTan[x]]]`
Out[·]:= 1

Taking *f* as the value of `ComplexityFunction` we can convert our expression to the same form as we obtained by using `FunctionExpand`:

In[·]:= `FullSimplify[Sin[2*ArcTan[x]], ComplexityFunction -> f]`
Out[·]:= $\frac{2x}{1+x^2}$

or

In[·]:= `FullSimplify[Sin[2*ArcTan[x]], ComplexityFunction ->`
`(Count[#1, Sin, Infinity, Heads -> True] & )]`
Out[·]:= $\frac{2x}{1+x^2}$

Here are some more examples:

In[·]:= `FullSimplify[Sin[x]^3, ComplexityFunction -> f]`
Out[·]:= $-\frac{1}{8} I E^{-3Ix} (1 - E^{Ix})^3 (1 + E^{Ix})^3$

In[·]:= `FullSimplify[Sin[x]^3, ComplexityFunction ->`
        `(30*f[#1] + LeafCount[#1] & )]`
Out[·]:= $\frac{1}{8} I E^{-3Ix} (-1 + E^{2Ix})^3$

Note that in spite of its name, `FunctionExpand` can also convert certain constants to their numerical values:

In[·]:= `FunctionExpand[GoldenRatio]`
Out[·]:= $\frac{1}{2}(1 + \sqrt{5})$

In[·]:= `FunctionExpand[GoldenAngle]`
Out[·]:= $(3 - \sqrt{5})\pi$

1.3 Assumptions in Mathematica®

It is possible to specify arbitrary symbolic assumptions about variables in the Wolfram Language. `Assumptions` is an option for functions like `Simplify`, `Refine` and `Integrate` and can take the form of equations, inequalities and domains. This option should not be confused with `$Assumptions`, which is a default setting for `Assumptions` and which can be modified temporarily by the function `Assuming`.

Sometimes it is necessary to specify the domain to which the values of the symbols belong. Let us compare the following outputs:

In[·]:= `Simplify[Abs[x^2]]`
Out[·]:= $Abs[x]^2$

In[·]:= `Assuming[Element[x, Reals], Simplify[Abs[x^2]]]`
Out[·]:= x^2

In[·]:= `Simplify[Abs[Cos[x]] <= 1]`
Out[·]:= $Abs[Cos[x]] \le 1$

In[·]:= `Assuming[Element[x, Reals], Simplify[`
        `Abs[Cos[x]] <= 1]]`
Out[·]:= `True`

In[·]:= `Reduce[Abs[x] == -x, Element[x, Reals]]`
Out[·]:= $x \le 0$

Note that the function `Refine`, unlike `Simplify`, just applies assumptions but does not attempt to make any other simplifications, not related to the assumptions:

In[·]:= Refine[Sqrt[x^2], Assumptions -> {Element[x, Reals]}]
Out[·]:= Abs[x]

In[·]:= Assuming[x > 0, Refine[Sqrt[x^2] +
          Cos[x]^2 + Sin[x]^2]]
Out[·]:= x + Cos[x]^2 + Sin[x]^2

In[·]:= Simplify[Sqrt[x^2] + Cos[x]^2 + Sin[x]^2,
          Assumptions -> {x > 0}]
Out[·]:= 1 + x

By default Mathematica® makes simplifications which are valid for complex numbers. If the user wants some simplifications which require additional assumptions which do not hold in general, one has to use the Assumptions mechanism. This is, for example, why Mathematica® does not give the answer True to the following identity:

In[·]:= Simplify[Sqrt[x*y] == Sqrt[x]*Sqrt[y]]
Out[·]:= $\sqrt{xy}$ == $\sqrt{x}$ $\sqrt{y}$

However,

In[·]:= Assuming[x >= 0 && y >= 0, Simplify[
          Sqrt[x*y] == Sqrt[x]*Sqrt[y]]]
Out[·]:= True

The alternative form, which is more useful, is

In[·]:= Simplify[Sqrt[x]*Sqrt[y], Assumptions ->
          {x >= 0 && y >= 0}]
Out[·]:= $\sqrt{xy}$

Note that the assumption $x > 0$, or any assumption that uses an inequality sign, automatically implies that x is real.

One can also use a global variable $Assumptions to define certain assumptions that will hold throughout the session. For example,

In[·]:= $Assumptions = {Element[_, Reals]};
In[·]:= $Assumptions
Out[·]:= {_ ∈ $\mathbb{R}$}

This is a quick way to make the assumption that everything (this is what the pattern _ stands for) in any expression is real, for example,

In[·]:= Simplify[Re[x*y]]
Out[·]:= x y

(of course, we could have used the more limited assumption that the symbols x and y represent real numbers).

To remove the global assumptions we should use

In[·]:= $Assumptions = True

and we can see that the following expression above will then not be simplified:

In[·]:= Simplify[Re[x*y]]
Out[·]:= Re[x y]

We can also locally change assumptions:

In[·]:= Block[{$Assumptions = a > 0}, Refine[Sqrt[a^2]]]
Out[·]:= a

The function Assuming is effectively equivalent to this block construction.

There exists one exception to the discussion above: the function PowerExpand. This function, without any assumptions, will perform a transformation of powers in the expression, which might not be true in general. For example,

In[·]:= PowerExpand[Sqrt[x*y]]
Out[·]:= $\sqrt{x}\,\sqrt{y}$

As we know, this only holds if x and y are both non-negative. The reason for that behavior is simply practical convenience. If PowerExpand is used with Assumptions, then it behaves like any other Mathematica® function; in other words, it only performs expansions that are valid under the given assumptions. For example, the following expansion is valid only for $x \geq 0$:

In[·]:= PowerExpand[Sqrt[x^2]]
Out[·]:= x

Suppose we want to add the assumption $x < 0$. Then PowerExpand gives the correct answer:

In[·]:= PowerExpand[Sqrt[x^2], Assumptions -> x < 0]
Out[·]:= −x

Note that this is the only case in which Assuming and Assumptions are not equivalent. For example, in the following expression PowerExpand ignores conditions in the function Assuming and gives an incorrect answer:

In[·]:= Assuming[x < 0, PowerExpand[Sqrt[x^2]]]
Out[·]:= x

Therefore, one needs to be aware of the fact that assumptions in PowerExpand must be given by means of the Assumptions option and any other way (even the global setting with $Assumptions) will be ignored.

1.4 Quantifiers

Quantifiers play a big role in formulating axioms and theorems in pure mathematics. In algorithmic (computer) mathematics, they can also be used to solve non-trivial problems, by means of an algorithm called "quantifier elimination".

In Mathematica® quantifiers are expressed by functions ForAll and Exists. They can be entered with the help of the Writing Assistant palette.

In[·]:= ForAll[x, Element[x, Reals], x > 0]
Out[·]:= $\forall_{x,x\in\mathbb{R}} x > 0$

In[·]:= $\forall_{x,x\in\mathbb{R}} x > 0$
Out[·]:= $\forall_{x,x\in\mathbb{R}} x > 0$

In[·]:= Exists[x, Element[x, Reals], x > 0]
Out[·]:= $\exists_{x,x\in\mathbb{R}} x > 0$

By default Mathematica® does not apply any rules to quantified expressions, and therefore we have the same expression as the input and the output.

Applying Reduce or Resolve (and also FullSimplify) to an expression with quantifiers will cause Mathematica® to use a famous algorithm called "quantifier elimination", which (in certain situations) will return an equivalent expression without any quantifiers and without the associated bound variables (variables to which the quantifiers apply). The free variables (the ones that are not quantified) will remain. For instance, in the following example there is one bound variable x and a free variable a. After quantifier elimination an explicit condition on the parameter a is obtained:

In[·]:= Resolve[ForAll[x, x^2 + a > 0], Reals]
Out[·]:= a > 0

If there are no free variables then the result should be either True or False.

The difference between Resolve and Reduce is that Reduce will often return a more "reduced" expression, but at the cost of longer computation. In simple cases both functions will return the same answer:

In[·]:= Resolve[ForAll[x, Element[x, Reals], x > 0]]
Out[·]:= False

In[·]:= Reduce[ForAll[x, Element[x, Reals], x > 0]]
Out[·]:= False

In[·]:= Resolve[Exists[x, Element[x, Reals], x^2 < 0]]
Out[·]:= False

In[·]:= Reduce[Exists[x, Element[x, Reals], x^2 < 0]]
Out[·]:= False

One can verify the triangle inequality as follows:

In[·]:= Reduce[ForAll[{a, b}, Abs[a] + Abs[b] >= Abs[a + b]]]
Out[·]:= True

Let us try something more complicated. We ask: "what is the condition on b and c which will make the quadratic below positive for all values of x?"

In[·]:= Resolve[ForAll[x, Element[x, Reals],
x^2 + b*x + c > 0]]
Out[·]:= $b \in \mathbb{R}$ && $b^2 - 4c < 0$ && $c > 0$

This is an example where we can see that Reduce gives a more "reduced" answer than Resolve:

In[·]:= Reduce[ForAll[x, Element[x, Reals],
x^2 + b*x + c > 0]]
Out[·]:= $c > 0$ && $-2\sqrt{c} < b < 2\sqrt{c}$

Here is another interesting problem: we ask Mathematica® for the conditions when the quadratic $ax^2 + bx + c$ has two equal roots:

In[·]:= Resolve[ForAll[{x, y}, a*x^2 + b*x + c == 0 &&
a*y^2 + b*y + c == 0, x == y], {a, b, c}]
Out[·]:= $(a = 0 \&\& b \neq 0) \,||\, (a = 0 \&\& c \neq 0) \,||\, \left(a \neq 0 \&\& c = \dfrac{b^2}{4a} \right)$

The next example is more constructive. We want to determine real values of a for which the equation $x^4 - ax + 1 = 0$ has a real root:

In[·]:= ToRadicals[Reduce[Exists[x, Element[x, Reals],
x^4 - a*x + 1 == 0], x]]
Out[·]:= (Re[a] <= -(4/3^(3/4)) && Im[a] == 0) ||
(Re[a] >= 4/3^(3/4) && Im[a] == 0)

In[·]:= ToRadicals[Reduce[Exists[x, Element[x, Reals],
x^4 - a*x + 1 == 0], x, Reals]]
Out[·]:= a <= -(4/3^(3/4)) || a >= 4/3^(3/4)

Unfortunately the quantifier elimination algorithm, although surprising and beautiful, has a very high complexity, which means that if we try to use it on expressions with a larger number of free parameters, it will not finish in a reasonable time even on a very fast computer.

It is important to know that Mathematica® assumes that the symbols are complex numbers (unless they appear in an inequality) if no information about the domain is given. This is demonstrated by the following example, where Mathematica® assumed that x is a complex number:

In[·]:= Reduce[Exists[x, x^2 == -1]]
Out[·]:= True

If we want x to be real, we have to explicitly specify this:

In[·]:= Reduce[Exists[x, Element[x, Reals], x^2 == -1]]
Out[·]:= False

There are certain situations where we may get a surprising answer, for example,

In[·]:= Reduce[Exists[x, Element[x, Complexes], x^2 < 0]]
Out[·]:= False

This looks like a mistake, since of course there are complex numbers whose square is negative. However (for reasons connected with efficiency) the function Reduce by default assumes that anything that appears in an inequality is real. So if we really want to work with complex numbers we must inform Reduce about that explicitly by specifying Complexes as the domain of Reduce (it is not enough just to use Complexes inside the quantifier):

In[·]:= Reduce[Exists[x, x^2 < 0], Complexes]
Out[·]:= True

We state this here only for completeness, because we will be generally working with real numbers and such issues will not arise.

1.5 Complex numbers

There is a subject called Complex Analysis, which deals with functions of complex numbers. In this book we shall be primarily concerned with real analysis (although we will sometimes digress into complex topics) but we will begin with complex numbers.

In Wolfram Language we can work with both explicit complex numbers and symbolic complex variables. There are many built-in functions to work with complex numbers (e. g., to evaluate the real and imaginary part, the absolute value and the argument, to compute the complex conjugate of a given complex number, to convert to different forms of a complex number and so on).

When dealing with complex numbers, one of the most useful functions is ComplexExpand. ComplexExpand can work with numerical complex expressions, which it arranges in the standard form for representing complex numbers (the real part + the imaginary part):

In[·]:= ComplexExpand[1/(1 + I)]
Out[·]:= $\frac{1}{2} - \frac{I}{2}$

ComplexExpand can also work with symbolic complex variables. By default, without the second argument in ComplexExpand, all variables are assumed to be real, whereas if we specify that one or more variables are complex in the second argument, then the output is different:

In[·]:= `ComplexExpand[a + I*a + I*b]`
Out[·]:= `a + I (a + b)`

In[·]:= `ComplexExpand[a + b, b]`
Out[·]:= `a + I Im[b] + Re[b]`

In[·]:= `ComplexExpand[a + b, {a, b}]`
Out[·]:= `I (Im[a] + Im[b]) + Re[a] + Re[b]`

If an expression is real, then `ComplexExpand` acts just like `Expand`:

In[·]:= `ComplexExpand[(a + b)^3]`
Out[·]:= $a^3 + 3\,a^2 b + 3\,a b^2 + b^3$

If the expression is complex, then it will be arranged in the standard form:

In[·]:= `ComplexExpand[(a + I*b)^3]`
Out[·]:= $a^3 - 3\,a b^2 + I(3\,a^2 b - b^3)$

There is an option `TargetFunctions` of the function `ComplexExpand`. With this option `ComplexExpand` will try to give the result in terms of functions that are specified by the user from the list `Re, Im, Abs, Arg, Conjugate, Sign`. By default `Re` and `Im` are used. This is illustrated by the following examples, if the expression contains a complex symbol:

In[·]:= `ComplexExpand[a + b*I, a]`
Out[·]:= `I (b + Im[a]) + Re[a]`

In[·]:= `ComplexExpand[a + b*I, a, TargetFunctions ->`
`        {Arg, Abs}]`
Out[·]:= `Abs[a] Cos[Arg[a]] + I (b + Abs[a] Sin[Arg[a]])`

1.6 Real numbers

Rigorous courses of analysis (or any other branch of mathematics) begin with certain statements (called axioms) about undefined "primitive objects", which are assumed to be true (because one always has to begin somewhere). Then everything else is proved by using these statements and basic rules of logic (which, we assume, everyone is familiar with). However, the choice of both the "primitive objects" and the "axioms" is not unique. This is true even if we want to make the number of axioms as small as possible or as "basic" level as possible.

It is commonly agreed that the most basic level is that of set theory, and one can indeed set up the foundations of mathematics (and hence also of analysis) beginning with axioms of set theory [15]. There are, in fact, several such "axiomatic set theories" (e. g., the Zermelo–Fraenkel set theory). Each of them allows one to construct the natural numbers beginning with sets. Once natural numbers have been constructed, Peano axioms, which will be discussed in Section 1.11, then become theorems, and one can then construct the rationals, and eventually the real numbers.

Another possible approach is to start with the natural numbers (satisfying the Peano axioms) and then construct the rationals, reals and complexes. To do this we still need to use the language of set theory, but we can treat it as part of "logic".

Finally, we can take the quickest approach but also one that makes the most assumptions and introduces axioms that uniquely define the real numbers. The integers and the rationals are then defined as special kinds of real numbers.

Here we will take the last axiomatic approach to real numbers but we shall omit the axioms, assuming that the reader is already familiar with them (see, for example, [14]). In other words, we will assume that there exists a set $\mathbb{R}$ with three binary operations +, · and > such that $\mathbb{R}$ is a complete (commutative) ordered field. The meaning of "complete" will be explained later.

Mathematica® has a number of built-in "rules" which effectively implement these axioms for both complex and real numbers (recall that by default Mathematica® treats all variables as complex numbers).

Certain properties of complex numbers are carried out automatically on evaluation. For example, Mathematica® by default assumes that both addition and multiplication are commutative and associative so it immediately converts $b + a$ to $a + b$, and $b\,a$ to $a\,b$ (i. e., variables are automatically arranged in the canonical lexicographic order). Similarly, the associative laws for addition and multiplication are used automatically; for instance

In[·]:= (a + b) + c
Out[·]:= a + b + c

However, the distributive law is not applied automatically:

In[·]:= a*(b + c)
Out[·]:= a (b + c)

If we want Mathematica® to apply it we need to use some other function, for example Expand or Distribute:

In[·]:= Expand[a*(b + c)]
Out[·]:= a b + a c

In[·]:= Distribute[a*(b + c)]
Out[·]:= a b + a c

The reason why Mathematica® applies certain basic operations automatically is due to the need for efficiency in computations. This need even forces Mathematica® to perform certain operations, like cancelations, without checking that no division by 0 is involved:

In[·]:= (x*(x - y))/(x - y)
Out[·]:= x

The question of mathematical correctness of such a cancelation is moot. One could argue that the very fact that we write $1/x$ implies that x is non-zero. However, automatic canceling of common factors can (very rarely) have surprising consequences and even lead to wrong answers, so one should be aware of it. This problem is unavoidable since there is no algorithm that can decide in general if an expression is zero or not.

Mathematica® also has a built-in non-commutative multiplication, denoted by ** (though we shall not use it here):

```
In[·]:= Distribute[(a + b) ** (a + b)]
Out[·]:= a ** a + a ** b + b ** a + b ** b
```

One can check in Mathematica® that the method of "quantifier elimination" will verify the axioms of an ordered field for real numbers. For example, we use Resolve (or Reduce) and quantifiers to check that the distributive rule holds:

```
In[·]:= Resolve[ForAll[{x, y, z}, Element[{x, y, z}, Reals],
          x*(y + z) == x*y + x*z]]
Out[·]:= True
```

Similarly, other axioms can be checked, for instance

```
In[·]:= Resolve[ForAll[{x, y, z}, Element[{x, y, z}, Reals]
          && x <= y, x + z <= y + z]]
Out[·]:= True
```

```
In[·]:= Resolve[ForAll[{x, y, z}, z >= 0 &&
          x <= y, x*z <= y*z]]
Out[·]:= True
```

The following expression contains one quantified (bound) variable x and two free variables y and z:

```
In[·]:= Exists[{x}, Element[{x, y, z}, Reals],
          x + y == 0 && x + z == 0];
```

The statement says that there exists a (complex) number x which has two additive inverses, y and z. Eliminating quantifiers will tell us that y and z must be equal. This can be viewed as a proof that every number has a unique additive inverse.

```
In[·]:= Reduce[Exists[{x}, Element[{x, y, z}, Reals],
          x + y == 0 && x + z == 0]]
Out[·]:= z ∈ ℝ && y == z
```

The axioms of ordered field are insufficient to determine the real numbers uniquely. In other words, there exist many different ordered fields, which are not "isomorphic" to ℝ. (Two sets with additional mathematical structure are said to be isomorphic if there exists a bijective mapping between them, such that both the mapping and the inverse preserve the structure.) In fact, the set $\mathbb{Q}$ of rational numbers is also an ordered field, and the inclusion $\mathbb{Q} \to \mathbb{R}$ preserves all the structure, but, of course, it is not a bijection (the real numbers are uncountable). In fact, there is just one additional axiom (of completeness) which is needed to determine the real numbers uniquely, but we shall leave it for later (see Section 1.15).

One consequence of the fact that only real numbers can be ordered is that any function that involves ordering (e. g., functions that maximize or minimize) can be applied only to subsets of real numbers or to real-valued functions.

1.7 Infinities

As usual in analysis we extend the real line by two symbols ∞ and $-\infty$. In Mathematica® we can simply input them as `Infinity` (∞ in `TraditionalForm`) and `-Infinity` ($-\infty$), but it is useful to note that the full forms of these expressions are `DirectedInfinity[1]` and `DirectedInfinity[-1]`:

In[·]:= `FullForm[Infinity]`
Out[·]//FullForm= `DirectedInfinity[1]`

In[·]:= `DirectedInfinity[-1]`
Out[·]:= $-\infty$

In fact, Mathematica®'s function `DirectedInfinity` gives an "infinity" in the direction of each complex number z (for instance, `DirectedInfinity[I]` denotes infinity in the direction pointing up along the imaginary axis). This may be useful in certain problems of complex analysis, for instance

In[·]:= `Limit[E^z, z -> DirectedInfinity[-1]]`
Out[·]:= `0`

In[·]:= `Limit[E^z, z -> DirectedInfinity[1]]`
Out[·]:= ∞

There is also a non-directed complex infinity, `ComplexInfinity`, whose FullForm is `DirectedInfinity[ ]`:

In[·]:= `FullForm[ComplexInfinity]`
Out[·]//FullForm= `DirectedInfinity[]`

`ComplexInfinity` will play no role in this book but it will sometimes occur in the output of some computations.

There is also another symbol, `Indeterminate`, which is used whenever a certain operation cannot be performed:

In[·]:= `0/0`
    $\cdots$ Power: Infinite expression $\frac{1}{0}$ encountered.
    $\cdots$ Infinity: Indeterminate expression 0 ComplexInfinity encountered.
Out[·]:= `Indeterminate`

Certain arithmetical operations are defined on infinities. Those operations which are not defined return `Indeterminate`. Here we give a couple of examples of the ones that are defined:

In[·]:= `Infinity + Infinity`
Out[·]:= ∞

In[·]:= `Infinity*Infinity`
Out[·]:= ∞

In[·]:= `Infinity + 1`
Out[·]:= ∞

In[·]:= `2*Infinity`
Out[·]:= ∞

In[·]:= `1/Infinity`
Out[·]:= `0`

Examples of undefined operations are

In[·]:= `Infinity - Infinity`
··· `Infinity: Indeterminate expression` $-\infty + \infty$ `encountered.`
Out[·]:= `Indeterminate`

In[·]:= `Infinity/Infinity`
··· `Infinity: Indeterminate expression` $0\,\infty$ `encountered.`
Out[·]:= `Indeterminate`

In[·]:= `0*Infinity`
··· `Infinity: Indeterminate expression` $0\,\infty$ `encountered.`
Out[·]:= `Indeterminate`

Note also that

In[·]:= `1/0`
··· `Power: Infinite expression` $\frac{1}{0}$ `encountered.`
Out[·]:= `ComplexInfinity`

Whenever operations involving infinities are defined, they obey all the axioms of an ordered field:

In[·]:= `2 < Infinity`
Out[·]:= `True`

In[·]:= `-Infinity < -10`
Out[·]:= `True`

1.8 Numbers in different bases

Mathematica® allows one to write numbers in an arbitrary base.[1]

In[·]:= `BaseForm[8, 2]`
Out[·]//BaseForm= 1000_2

which is equivalent to

In[·]:= `1*2^3 + 0*2^2 + 0*2^1 + 0*2^0`
Out[·]:= 8

In[·]:= `%% // FullForm`
Out[·]//FullForm= 8

The function `IntegerDigits` gives a list of the digits:

In[·]:= `IntegerDigits[8, 2]`
Out[·]:= `{1, 0, 0, 0}`

One can convert back to base 10 using `FromDigits`, which constructs an integer from the list of its decimal digits:

In[·]:= `FromDigits[%, 2]`
Out[·]:= 8

The function `IntegerString` gives the digits as a string:

In[·]:= `IntegerString[8, 2]`
Out[·]:= 1000

Converting the string back to a base-10 number can be done using the same function:

In[·]:= `FromDigits[%, 2]`
Out[·]:= 8

1.9 Interval arithmetic

Interval arithmetic is a mathematical technique for estimating and controlling errors in approximate computations.[2] Its weakness is that it is a relatively slow approach to this problem. Mathematica has its own faster but less reliable "approximate version" of interval arithmetic, which it uses for error control in numerical computations with extended precision numbers (see, for example, the documentation for `Precision` and `Accuracy`). However, it also implements strict interval arithmetic, by means of the function `Interval`. `Interval` represents the range of values between a minimum value and a

---

1 https://reference.wolfram.com/language/ref/BaseForm.html
https://mathworld.wolfram.com/DecimalExpansion.html
2 https://en.wikipedia.org/wiki/Interval_arithmetic

maximum value. In Mathematica® one can perform certain arithmetic operations with them:

In[·]:= `Interval[{1, 2}] + Interval[{-1, 0}]`
Out[·]:= `Interval[{0, 2}]`

In[·]:= `Interval[{-1, 0}]*Interval[{-1, -1}]`
Out[·]:= `Interval[{0, 1}]`

In[·]:= `Interval[{-1, 1}]^2`
Out[·]:= `Interval[{0, 1}]`

Interval arithmetic can be used in computations as follows:

In[·]:= `x^2 + 3*x + 1 /. x -> Interval[{0.3, 0.4}]`
Out[·]:= `Interval[{1.99, 2.36}]`

In[·]:= `Sin[Infinity]`
Out[·]:= `Interval[{-1, 1}]`

1.10 Telescoping in Mathematica®

Normally Mathematica® evaluates sums and products, e. g.,

In[·]:= `Product[((i - 1)*(i + 2))/(i*(i + 1)), {i, 2, n}]`
Out[·]:= `(2 + n)/(3*n)`

Sometimes one prefers to keep the result unevaluated or partially evaluated. For example, in this case, writing out explicitly a few terms of the product, one notices that it "telescopes" (terms cancel):

In[·]:= `DisplayForm[RowBox[Table[FractionBox[RowBox[`
`    {ToString[i - 1], " ", ToString[i + 2]}], RowBox[`
`    {ToString[i], " ", ToString[i + 1]}]], {i, 2, 6}]]]`
Out[·]//DisplayForm= $\frac{1\times4}{2\times3} \frac{2\times5}{3\times4} \frac{3\times6}{4\times5} \frac{4\times7}{5\times6} \frac{5\times8}{6\times7}$

In Wolfram Language one can produce accurate-looking formulas like above by using low level formatting functions (involving the word "Box"). However, this approach is time-consuming and one can obtain results which are good enough in practice by special functions which prevent or delay the evaluation of parts of expressions. One way to do it is by means of the function `HoldForm`. In the example below we temporarily replace the `Head` "Times" with "times" to prevent evaluation:

In[·]:= `Product[times[i - 1, i + 2]/times[i, i + 1],`
`    {i, 2, 4}] /. times[x_, y_] :> HoldForm[x*y]`
Out[·]:= $\frac{(1\times4)(2\times5)(3\times6)}{(2\times3)(3\times4)(4\times5)}$

In version 10 of Mathematica® the function `Inactivate` was introduced, which allows one to temporarily inactivate chosen parts of expressions:

In[·]:= `Inactivate[Product[((i - 1)*(i + 2))/(i*(i + 1)),`
    `{i, 2, 4}], Times]`
Out[·]:= $\left((1 \times 4) \times \frac{1}{2 \times 3}\right)\left((2 \times 5) \times \frac{1}{3 \times 4}\right)\left((3 \times 6) \times \frac{1}{4 \times 5}\right)$

This approach allows us to inactivate and activate chosen parts of expressions:

In[·]:= `expr = Inactivate[2 + 2 + 3^2]`
Out[·]:= `2 + 2 + 3^2`

In[·]:= `Activate[expr, Plus]`
Out[·]:= `4 + 3^2`

In[·]:= `Activate[expr, Power]`
Out[·]:= `2 + 2 + 9`

In[·]:= `Activate[expr]`
Out[·]:= `13`

Another function with similar properties is `Defer`:

In[·]:= `expr1 = Defer[1 + 1]`
Out[·]:= `1 + 1`

Deferred output does not need to be explicitly activated. Instead, one can copy it, paste somewhere and evaluate as one would do with ordinary input:

In[·]:= `1 + 1`
Out[·]:= `2`

In[·]:= `FullForm[expr1]`
Out[·]//FullForm= `Defer[Plus[1, 1]]`

In[·]:= `expr1[[1]]`
Out[·]:= `2`

1.11 Integers and the Principle of Mathematical Induction

As we have already mentioned before, the most common axiomatic approach to analysis is based on constructing the real numbers after beginning with sets or with integers. For example, one can begin with the so called system of axioms of Peano ([14]).

Peano's axioms. There exists a set $\mathbb{N}$ with an element $1 \in \mathbb{N}$ and a function $S : \mathbb{N} \to \mathbb{N}$ (called the successor function) which satisfies the following axioms:

1. $\forall_{n, n\in\mathbb{N}} \exists_{S(n)} S(n) \in \mathbb{N}$ (each $n \in \mathbb{N}$ has a successor).
2. $S(n) \neq 1$ (1 is not the successor of any $n \in \mathbb{N}$).
3. $S(n) = S(m) \Leftrightarrow n = m$ (if m and n in $\mathbb{N}$ have the same successor, then $m = n$, i. e., two distinct elements in $\mathbb{N}$ cannot have the same successor).
4. If $A \subset \mathbb{N}$ has the properties

- 1 ∈ A and
- $n \in A \Rightarrow S(n) \in A$,

then $A = \mathbb{N}$.

The last axiom is called the *induction axiom*.

It turns out that by starting only with these axioms we can construct the set of real numbers satisfying all the axioms (including the completeness axiom below). Alternatively, if we start with the axioms of real numbers, we first define $S : \mathbb{R}_+ \to \mathbb{R}_+$ (where $\mathbb{R}_+ = \{x \in \mathbb{R} \mid x > 0\}$) by $S(x) = x + 1$. A subset A of the real numbers possessing the property in the hypothesis of axiom 4 is called an inductive subset of $\mathbb{R}$. For example, $\mathbb{R}_+$ is an inductive subset of $\mathbb{R}$. We now define $\mathbb{N}$ as the intersection of all inductive subsets of $\mathbb{R}_+$. It is easy to prove that $\mathbb{N}$ defined in this way satisfies all of Peano's axioms. In particular, axiom 4 now becomes a theorem (known as the "Principle of Mathematical Induction"), whose proof follows almost trivially from the definition. The result is extremely useful in proving theorems in which some statement is asserted to hold for all natural numbers (or possibly integers larger than some given integer).

The *Principle of Mathematical Induction* works as follows. Suppose we wish to prove that some sentence A_n, which depends on the integer n, holds for all $n \geq 1$. We consider the set $A = \{k \in \mathbb{N} \mid A_k \text{ is true}\}$. We first prove that $1 \in A$, i. e., A_1 is true. Finally we prove that $n \in A \Rightarrow n + 1 \in A$. Thus, A satisfies the hypothesis of axiom 4, hence it is equal to $\mathbb{N}$. Thus A_n is true for all $n \in \mathbb{N}$.

Let us consider two examples.

1.11.1 Example

Consider the finite sum

$$\sum_{k=1}^{n} k^3 = 1^3 + 2^3 + \cdots + n^3.$$

Let us ask Mathematica® to find a closed form formula for this sum:

In[·]:= Sum[i^3, {i, 1, n}]

Out[·]:= $\dfrac{1}{4} n^2 (1 + n)^2$

Now that we know the answer let us try to prove it by induction. Let A_1 state that $1^3 = 1^2 2^2/4$, which is true. Now suppose A_n is true for some $n \in \mathbb{N}$. Then

$$\sum_{k=1}^{n+1} k^3 = \sum_{k=1}^{n} k^3 + (n + 1)^3 = \frac{1}{4} n^2 (n + 1)^2 + (n + 1)^3$$

$$= \frac{1}{4}(n + 1)^2(n^2 + 4n + 4) = \frac{1}{4}(n + 1)^2(n + 2)^2.$$

Hence, A_{n+1} is also true.

Next we give a different example, which illustrates how Mathematica® can be used to suggest a solution to a problem, which it alone cannot solve.

1.11.2 Example

Let p_n denote the n-th prime number. Show that $\forall_{n,\,n\in\mathbb{N}\,\&\&\,n\geq 12}\, p_n > 3n$.

Before we come to the proof (where we will use induction), let us try to see how we could discover this statement with the help of Mathematica®. Mathematica® has a built-in function Prime[n], which gives the n-th prime number.

In[·]:= primes = Table[Prime[i], {i, 1, 10}]
Out[·]:= {2, 3, 5, 7, 11, 13, 17, 19, 23, 29}

The function Prime is Listable in Mathematica®:

In[·]:= Attributes[Prime]
Out[·]:= {Listable, Protected, ReadProtected}

so we can easily obtain the list of prime numbers:

In[·]:= Prime[Range[10]]
Out[·]:= {2, 3, 5, 7, 11, 13, 17, 19, 23, 29}

To check if the theorem is true for small integers n we can plot the functions Prime[n] and $3n$ together:

In[·]:= DiscretePlot[{Prime[n], 3*n}, {n, 1, 50}, PlotMarkers
           -> Automatic, Filling -> None]

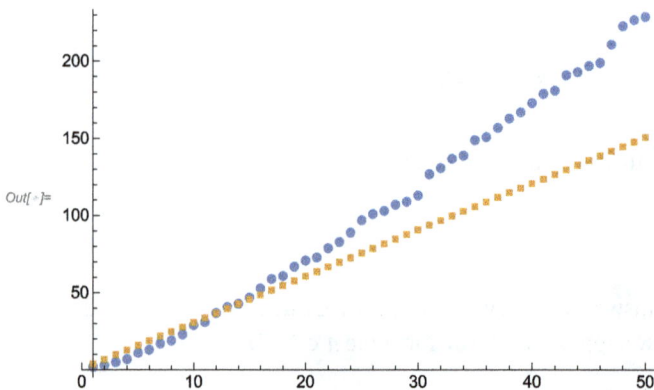

Figure 1.2

We see that once the graph of the function Prime[n] (discs) crosses that of $3n$ (squares), it seems to stay above the latter, although the rate of growth of Prime[n] is not always greater than that of $3n$ (which is three). To see how Prime[n] increases, we will look at the graph of the function that measures the differences between successive primes:

In[·]:= Differences[primes]
Out[·]:= {1, 2, 2, 4, 2, 4, 2, 4, 6}

In[·]:= DiscretePlot[Prime[i + 1] - Prime[i],
            {i, 1, 50, 1}, Joined -> True]

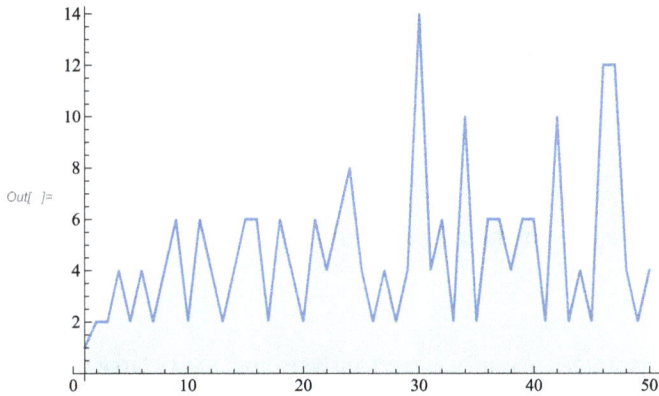

Figure 1.3

We see (what is actually quite obvious) that the differences must be at least 2. The question of whether there are infinitely many pairs of primes differing by 2 is a famous unsolved problem (recently there have been some remarkable results[3]). We can also see that two successive "jumps" of size 2 occur once $(3, 5, 7)$ and it is easy to prove that this can never happen again. So now it should be easy to complete a proof: basically the function $3n$ goes up in steps of 3, the function Prime can go up by 2 but in the next step it has to go up by at least 4. Once it is ahead by at least 2, it can never fall behind.

Let us now give a formal inductive proof. We first check that the statement is true for $n = 12$. Indeed,

In[·]:= Prime[12]
Out[·]:= 37

In[·]:= 12*3
Out[·]:= 36

We now assume that the result holds for some $n \geq 12$, that is, $p_n > 3n$, and we will try to show that $p_{n+1} > 3n + 3$. There are two possibilities. Either $p_{n+1} - p_n > 2$, in which case $p_{n+1} - p_n \geq 4$ (the difference must be even) and the inductive step follows, or $p_{n+1} - p_n = 2$.

3 https://www.nature.com/news/first-proof-that-infinitely-many-prime-numbers-come-in-pairs-1.12989

But then, by the inductive hypothesis $p_{n+1} = p_n + 2 > 3n + 2$, hence $p_{n+1} \geq 3n + 3$. But, of course, we cannot have $p_{n+1} = 3n + 3$, hence $p_{n+1} > 3n + 3$ and the inductive step is complete.

The function PrimePi returns the number of prime numbers that are less than a given number:

In[·]:= PrimePi[29.4]
Out[·]:= 10

In standard form it is denoted by π:

In[·]:= $\pi(\pi)$
Out[·]:= 2

In[·]:= PrimePi[-5]
Out[·]:= 0

We can also check whether a given number is prime or not:

In[·]:= PrimeQ[13]
Out[·]:= True

The function PrimeQ is guaranteed to return correct answers until a very large number but it is not known if it is always correct. Mathematica® includes a package called Primality Proving Package (PrimalityProving/tutorial/PrimalityProving), in which a function ProvablePrimeQ is defined. This function is guaranteed to always return correct answers but is much slower than PrimeQ.

Let us now generalize our example. First, let us compare p_n with $5n$.

In[·]:= DiscretePlot[{Prime[n], 5*n}, {n, 1, 100}]

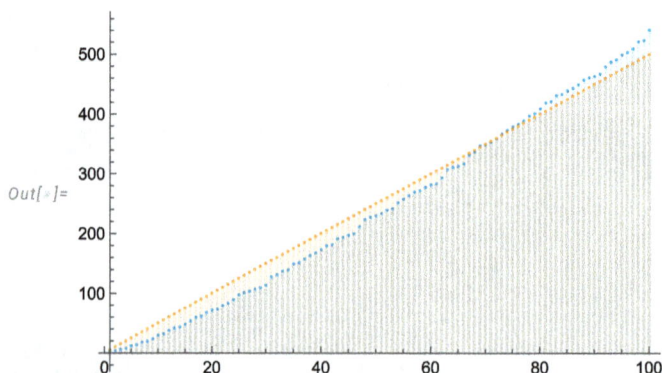

Figure 1.4

In[·]:= DiscretePlot[{Prime[n]/(5*n), 1}, {n, 1, 100},
        PlotStyle -> {Red, Green}, Filling -> None]

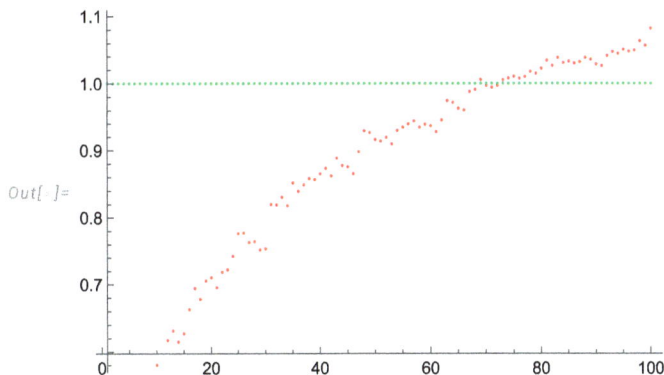

Out[ ]=

Figure 1.5

In[·]:= `n = 1; While[Prime[n] <= 5*n, n++]; n`
Out[·]:= 69

We could have also computed this using a local variable *n*:

In[·]:= `Block[{n = 1}, While[Prime[n]/(5*n) < 1, n++]; n]`
Out[·]:= 69

or similarly using `Module`:

In[·]:= `Module[{n = 1}, While[Prime[n]/(5*n) < 1, n++]; n]`
Out[·]:= 69

or

In[·]:= `Catch[Table[If[Prime[n]/(5*n) > 1, Throw[n]],`
`        {n, 1, 100}]]`
Out[·]:= 69

   We will digress here to discuss an important programming technique known as lo-
calization of variables. In Wolfram Language it is possible to program without using any
assignments to variables – this approach is known as functional programming. Most
users, however, prefer to use a mixed approach – with both functions and assignments
(e. g., when defining a named function, e. g., `f[x_]:=x^2`). When one uses assignments
to global variables one has to remember to use `Clear` or `Remove` before using a name to
which a value has been assigned. Forgetting to do so is one of the most common sources
of problems when programming in Wolfram Language. One way to avoid the problems
is by "localizing" variables. In Wolfram Language there are three functions that do this:
`Block`, `Module` and `With`. Let us consider the first two in more detail, and, in particular,
the difference between them. In many examples the difference is not noticed but they
localize variables in essentially different ways. `Block` temporarily changes the value of
a localized variable inside `Block` and then restores the old value after exiting `Block`.

In[·]:= n = 3
Out[·]:= 3

In[·]:= Block[{n = 7}, n + 1]
Out[·]:= 8

In[·]:= n
Out[·]:= 3

In the case of Module the variables are renamed so that their names do not conflict with any other names. Module creates a new variable, which can be seen using the function Trace:

In[·]:= n = 3
Out[·]:= 3

In[·]:= Module[{n = 7}, n + 1]
Out[·]:= 8

In[·]:= Trace[Module[{n = 7}, n + 1]]
Out[·]:= {Module[{n = 7}, n + 1], {n$94025 = 7, 7},
         {{n$94025, 7}, 7 + 1, 8}, 8}

In[·]:= ?n

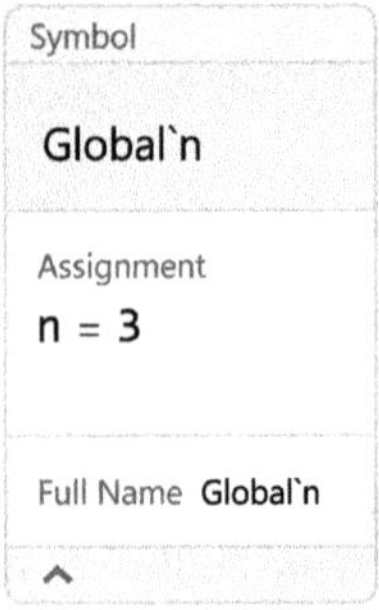

Figure 1.6

In[·]:= ?n$94025

Figure 1.7

Another helpful example is the following one. A global variable k is defined in terms of another global variable x. Inside `Block` one can (temporarily) change the value of k, by assigning a local value to x. Inside `Module` assigning a local value to x has no effect on k. Whether this is a good or a bad feature depends on the situation.

In[·]:= `k := x`

In[·]:= `x = 1; k`
Out[·]:= `1`

In[·]:= `Block[{x = 2}, k]`
Out[·]:= `2`

In[·]:= `k`
Out[·]:= `1`

In[·]:= `Module[{x = 2}, k]`
Out[·]:= `1`

In[·]:= `k`
Out[·]:= `1`

One more simple construction that localizes variables is `With`. Note that, unlike in `Module` and `Block`, all local variables in `With` must be initialized (or given values) so one cannot use assignments to local variables inside `With`. The function `With` resembles a substitution.

In[·]:= `With[{x = 2}, x + 2]`
Out[·]:= `4`

In[·]:= `Clear[v]; g[x_List] := With[{u = Length[x], v = u + 1}, v]`

In[·]:= `g[{1, 2, 3}]`
Out[·]:= `1 + u`

In[·]:= `v`
Out[·]:= `v`

We will obtain exactly the same result if we replace `With` with `Module`:

In[·]:= `Clear[v]; h[x_List] := Module[{u = Length[x], v = u + 1}, v]`

In[·]:= `h[{1, 2, 3}]`
Out[·]:= `1 + u`

In[·]:= `v`
Out[·]:= `v`

In[·]:= `u = 5`
Out[·]:= `5`

In[·]:= Clear[v]; h[x_List] := Module[{u = Length[x], v = u + 1}, v]

In[·]:= h[{1, 2, 3}]
Out[·]:= 6

In[·]:= v
Out[·]:= v

However, with Module we can also do this:

In[·]:= f[x_List] := Module[{u = Length[x], v}, v = u + 1]

In[·]:= f[{1, 2, 3}]
Out[·]:= 4

In[·]:= v
Out[·]:= v

In[·]:= Clear[u, v]; h[x_List] := Block[{u = Length[x], v = u + 1}, v]

In[·]:= h[{1, 2, 3}]
Out[·]:= 4

In[·]:= v
Out[·]:= v

Returning to our example, we observe that p_n first overtakes $5n$ for $n = 69$, but then falls behind and overtakes it again for $n = 73$. After that, it never again falls behind.

In[·]:= Table[Prime[n] - 5*n, {n, 69, 80}]
Out[·]:= {2, -1, -2, -1, 2, 3, 4, 3, 4, 7, 6, 9}

This can be proved by induction in a way analogous to the case of $3n$, but the proof is longer and we leave it as an exercise to the reader. Note that we can use DiscreteLimit to show that p_n will eventually overtake $5n$:

In[·]:= Clear[n]; DiscreteLimit[Prime[n]/(5*n), n -> Infinity]
Out[·]:= ∞

On the other hand, p_n increases more slowly than n^2:

In[·]:= DiscreteLimit[Prime[n]/n^2, n -> Infinity]
Out[·]:= 0

Note that Limit will not work above because Prime is defined only for positive integer arguments and Limit uses analytic methods:

In[·]:= Limit[Prime[n]/n^2, n -> Infinity]
Out[·]:= Limit[Prime[n]/n^2, n -> Infinity]

DiscreteLimit can compute limits of some sequences that cannot be computed by Limit.

The reason why Mathematica® is able to compute the limits is because it "knows" the "Prime Number Theorem".[4] In mathematics, the prime number theorem (PNT) describes the asymptotic distribution of the prime numbers among the positive integers. It formalizes the intuitive idea that primes become less common as they become larger by precisely quantifying the rate at which this occurs.

In[·]:= `f[x_] := x/Log[x]`

In[·]:= `Plot[{f[x], PrimePi[x]}, {x, 2, Infinity},`
`        PlotStyle -> {Red, Blue}]`

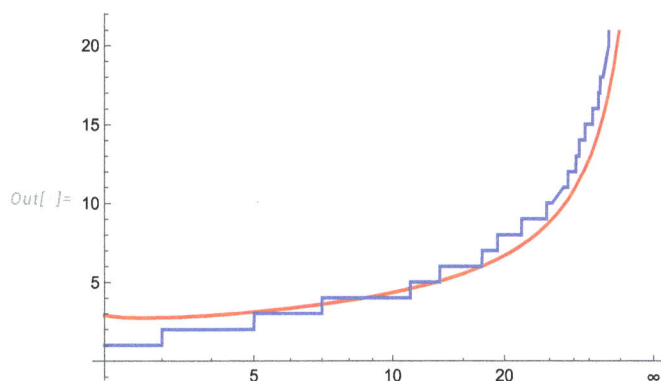

Figure 1.8

Here the function `PrimePi[x]` gives the number of primes $\pi(x)$ less than or equal to x.

In[·]:= `Limit[PrimePi[x]/f[x], x -> Infinity]`
Out[·]:= 1

or, alternatively,

In[·]:= `Asymptotic[PrimePi[x], x -> Infinity]`
Out[·]:= `x/Log[x]`

From this one can deduce that

In[·]:= `DiscreteLimit[Prime[n]/(n*Log[n]),`
`        n -> Infinity]`
Out[·]:= 1

Note that from the Prime Number Theorem it follows that p_n will eventually overtake any linear function but finding the exact value of n for which it occurs is in general a difficult problem.

---

4 https://en.wikipedia.org/wiki/Prime_number_theorem

1.12 Algebraic equations and algebraic numbers

Although real analysis can be studied without any mention of complex numbers, it is not a good approach when one is interested in the way computer programs do analysis. That is because many algorithms, even those that compute ultimately real quantities, make use of complex numbers. Probably the most profound example of this phenomenon is the Risch algorithm for computing indefinite integrals, which we will discuss later on in this book. The most famous classic example is solving polynomial equations (this example, of course, really belongs to algebra rather than analysis but we shall not engage in such pedantry here). Mathematica® by default assumes, unless specified otherwise, that one is working over the field of complex numbers when solving algebraic equations. As mentioned in Section 1.2, in order to obtain a real answer one needs to specify additionally the third argument, the domain Reals.

Instead of the function Solve we will often use the function Reduce, which can solve a variety of equations and inequalities but gives answers in a different form. Both functions have their advantages, but to understand them better the reader should consult the documentation.

Now let us now try an equation with a parameter:

In[·]:= sols = Solve[x^3 - a == 0, x]
Out[·]:= {{x -> a$^{1/3}$}, {x -> -(-1)$^{1/3}$ a$^{1/3}$}, {x -> (-1)$^{2/3}$ a$^{1/3}$}}

It is easy to prove that a cubic polynomial equation must possess at least one real root. Mathematica® can verify that they satisfy the original equation:

In[·]:= Simplify[x^3 - a /. sols]
Out[·]:= {0, 0, 0}

Now, suppose we only wanted to know the real root. We can ask Mathematica® to find it (using either Solve or Reduce):

In[·]:= Reduce[x^3 - a == 0, x, Reals]
Out[·]:= x == Root[-a + #1^3 & , 1]

This strange looking answer tells us that there is exactly one real root. To express it, Mathematica® does not use radicals but the "root object" Root. It is a symbolic expression beginning with the head Root, whose second argument is a number from 1 to the degree of the equation (in this case 1). You should think of these "root objects" as symbolic expressions like $\sqrt{2}$ or $\sqrt{a}$. If we replace a by a number, Mathematica® can compute this expression with an arbitrary precision. For example

In[·]:= N[Root[-2 + #1^3 & , 1], 3]
Out[·]:= 1.26

This confirms that the root found by Mathematica® above is indeed real. We can also check that the other roots are not:

In[·]:= N[Reduce[x^3 - 2 == 0, x], 3]
Out[·]:= x == -0.630 - 1.091 I || x == 1.26 ||
          x == -0.630 + 1.091 I

One more useful function, the function Roots (not Root), is used to obtain a disjunction (factorization) of equations which represent the roots of a polynomial equation:

In[·]:= Roots[(x - 1)*(x - 2)^2*(x - 3)^2 == 0, x]
Out[·]:= x == 3 || x == 3 || x == 2 || x == 2 || x == 1

Compare this with the output given by Reduce, where the multiple roots are not taken into account:

In[·]:= Reduce[(x - 1)*(x - 2)^2*(x - 3)^3 == 0, x]
Out[·]:= x == 1 || x == 2 || x == 3

One can choose in NSolve how many roots of the equation we want to find:

In[·]:= NSolve[x^111 + x + 1 == 0, x, MaxRoots -> 2]
Out[·]:= {{x -> -0.969148}, {x -> -0.972864 - 0.0460068 I}}

The function ToRules takes logical combinations of equations, in the form generated by Roots and Reduce, and converts them to lists of rules of the form produced by Solve. For example,

In[·]:= ToRules[x == 1 || x == 2]
Out[·]:= Sequence[{x -> 1}, {x -> 2}]

In[·]:= {ToRules[x == 1 || x == 2]}
Out[·]:= {{x -> 1}, {x -> 2}}

Let us consider another example (below we give outputs in Mathematica®11.3):

In[·]:= rts = x /. Solve[x^3 - 15*x + 2 == 0, x]

$$Out[\cdot]:= \{ \frac{5}{(-1 + 2I\sqrt{31})^{1/3}} + (-1 + 2I\sqrt{31})^{1/3},$$

$$-\frac{5(1 + I\sqrt{3})}{2(-1 + 2I\sqrt{31})^{1/3}} - \frac{1}{2}(1 - I\sqrt{3})(-1 + 2I\sqrt{31})^{1/3},$$

$$-\frac{5(1 - I\sqrt{3})}{2(-1 + 2I\sqrt{31})^{1/3}} - \frac{1}{2}(1 + I\sqrt{3})(-1 + 2I\sqrt{31})^{1/3} \}$$

The reason why this may appear strange is that actually all the roots of this equation are real, as can be shown by asking Mathematica®:

In[·]:= FullSimplify[(Element[#1, Reals] &) /@ rts]
Out[·]:= {True, True, True}

The problem here is that, although the roots are real, there is no way to express them in terms of radicals without using complex *i*. One can, of course, show that the roots are real by evaluating them numerically, or by using the function ComplexExpand:

In[·]:= Simplify[ComplexExpand[rts]]

Out[·]:= $\{\sqrt{5}\left(\cos[\frac{1}{3}\text{ArcTan}[2\sqrt{31}]] + \sqrt{3}\sin[\frac{1}{3}\text{ArcTan}[2\sqrt{31}]]\right),$

$-2\sqrt{5}\cos[\frac{1}{3}\text{ArcTan}[2\sqrt{31}]],$

$\sqrt{5}\left(\cos[\frac{1}{3}\text{ArcTan}[2\sqrt{31}]] - \sqrt{3}\sin[\frac{1}{3}\text{ArcTan}[2\sqrt{31}]]\right)\}$

We have obtained expressions for all three roots that are clearly real but they are not expressed in terms of radicals. The famous theorems due to Abel and (more generally) to Galois state that the roots of arbitrary polynomial equations of degree greater or equal to five cannot be expressed in terms of radicals (like $\sqrt[5]{2+i}$). Mathematica® can still solve such equations, but expresses the solutions using Root:

In[·]:= x /. Solve[x^5 + 2*x^3 + x - 2 == 0, x]

Out[·]:= {Root[-2 + #1 + 2#1^3 + #1^5 &, 1],

Root[-2 + #1 + 2#1^3 + #1^5 &, 2], Root[-2 + #1 + 2#1^3 + #1^5 &, 3],

Root[-2 + #1 + 2#1^3 + #1^5 &, 4], Root[-2 + #1 + 2#1^3 + #1^5 &, 5]}

Both Root objects and radicals are examples of algebraic numbers.[5] Of course, some special higher degree equations can be solved in radicals (we again present the output in Mathematica®11.3):

In[·]:= sols = x /. Solve[x^5 + 2*x^3 + x - 1 == 0, x]

Out[·]:= $\{-(-1)^{1/3}, (-1)^{2/3}, \frac{1}{3}\left(1 - 5\left(\frac{2}{11 + 3\sqrt{69}}\right)^{1/3} + \left(\frac{1}{2}(11 + 3\sqrt{69})\right)^{1/3}\right),$

$\frac{1}{3} - \frac{1}{6}(1 + I\sqrt{3})\left(\frac{1}{2}(11 + 3\sqrt{69})\right)^{1/3} + \frac{5(1 - I\sqrt{3})}{3 \times 2^{2/3}(11 + 3\sqrt{69})^{1/3}},$

$\frac{1}{3} - \frac{1}{6}(1 - I\sqrt{3})\left(\frac{1}{2}(11 + 3\sqrt{69})\right)^{1/3} + \frac{5(1 + I\sqrt{3})}{3 \times 2^{2/3}(11 + 3\sqrt{69})^{1/3}}\}$

RootReduce will try to reduce the solutions to a simpler form or to convert the radicals to Root objects (except for the case of roots of degree 1 and 2):

In[·]:= RootReduce[sols]

Out[·]:= $\{\frac{1}{2}(-1 - I\sqrt{3}), \frac{1}{2}(-1 + I\sqrt{3}),$ Root[-1 + 2#1 $-$ #1^2 + #1^3 &, 1],

Root[-1 + 2#1 $-$ #1^2 + #1^3 &, 2], Root[-1 + 2#1 $-$ #1^2 + #1^3 &, 3]}

On the other hand, the inverse operation, ToRadicals, only works in special cases, e. g.,

In[·]:= ToRadicals[Root[5 - 4*#1^3 + #1^6 & , 6]]

Out[·]:= $(2 + I)^{1/3}$

Mathematica®'s Root objects represent complex algebraic numbers – that is, complex numbers which are solutions of polynomial equations with integer (or rational) coefficients. Complex numbers which are not solutions of any polynomial equations are called transcendental numbers.

---

5 http://reference.wolfram.com/language/ref/AlgebraicNumber.html

In[·]:= Element[Sqrt[2], Algebraics]
Out[·]:= True

Mathematica® knows that π and e are transcendental.

In[·]:= Element[Pi, Algebraics]
Out[·]:= False

In[·]:= Element[E, Algebraics]
Out[·]:= False

It is not known whether $\pi + e$ or $e\pi$ are transcendental:

In[·]:= FullSimplify[Element[E + Pi, Algebraics]]
Out[·]:= Element[E + Pi, Algebraics]

In[·]:= FullSimplify[Element[E*Pi, Algebraics]]
Out[·]:= Element[E*Pi, Algebraics]

However, it easy to prove that at least one of them must be transcendental.

The algebraic numbers form a field, that is, their sums, products and the inverses of non-zero algebraic numbers are themselves algebraic. Each algebraic number has the so called "minimum polynomial", that is, a monic polynomial of the smallest degree of which it is a root. For example,

In[·]:= Clear[x]; MinimalPolynomial[Sqrt[2], x]
Out[·]:= $-2 + x^2$

In[·]:= MinimalPolynomial[Sqrt[2] + Sqrt[3], x]
Out[·]:= $1 - 10 x^2 + x^4$

In[·]:= Element[Sqrt[2] + 3^(1/3), Algebraics]
Out[·]:= True

Mathematica® can find the minimal polynomial of this algebraic number:

In[·]:= InputForm[RootReduce[Sqrt[2] + 3^(1/3)]]
Out[·]//InputForm= Root[1 - 36*#1 + 12*#1^2 - 6*#1^3 - 6*#1^4
        + #1^6 & , 2, 0]

In[·]:= MinimalPolynomial[Sqrt[2] + 3^(1/3), x]
Out[·]:= $1 - 36 x + 12 x^2 - 6 x^3 - 6 x^4 + x^6$

Rational numbers are precisely the numbers whose minimal polynomials have degree 1.

Note that the first argument of a Root object is the minimal polynomial of the algebraic number it represents, written in the form of a pure function. In recent versions of Mathematica® Root objects are represented by a root sign together with a numerical approximation:

In[·]:= Solve[1 - 10*x + x^3 == 0, x]

Out[∘]= {{x → ③⊙ -3.21...}, {x → ③⊙ 0.100...}, {x → ③⊙ 3.11...}}

Figure 1.9

However, expressing the output in InputForm one can see the minimal polynomials and the ordering of the roots:

In[·]:= InputForm[%]
Out[·]//InputForm= {{x -> Root[1 - 10*#1 + #1^3 & , 1, 0]},
   {x -> Root[1 - 10*#1 + #1^3 & , 2, 0]},
   {x -> Root[1 - 10*#1 + #1^3 & , 3, 0]}}

For polynomials Mathematica® isolates the roots by constructing rectangles in the complex plane so that distinct roots are contained in distinct rectangles. Real roots always come before non-real ones. The second argument of the function Root is the number of the root in Mathematica®'s canonical ordering. (There are also two different methods of isolating roots. The third argument in Root, which is either 1 or 0, refers to the method used).

1.12.1 Example

Is the number

$$\sqrt{\sqrt{11} - 3} + \sqrt{\sqrt{11} + 5}$$

rational?

The function Element can immediately determine that the number is algebraic:

In[·]:= Element[Sqrt[Sqrt[11] - 3] +
   Sqrt[Sqrt[11] + 5], Algebraics]
Out[·]:= True

but not if it is rational:

In[·]:= Element[Sqrt[Sqrt[11] - 3] +
   Sqrt[Sqrt[11] + 5], Rationals]
Out[·]:= Element[Sqrt[-3 + Sqrt[11]] +
   Sqrt[5 + Sqrt[11]], Rationals]

RootReduce is needed to determine the answer:

In[·]:= Element[RootReduce[Sqrt[Sqrt[11] - 3] +
   Sqrt[Sqrt[11] + 5]], Rationals]
Out[·]:= False

In[·]:= FullSimplify[Element[Sqrt[Sqrt[11] - 3] +
   Sqrt[Sqrt[11] + 5], Rationals],
   TransformationFunctions -> {Automatic, RootReduce}]
Out[·]:= False

Equivalently, one could simply compute the minimal polynomial, which happens to have degree 8:

In[·]:= `MinimalPolynomial[Sqrt[Sqrt[11] - 3] +`
`        Sqrt[Sqrt[11] + 5], x]`
Out[·]:= $4096 - 512 x^2 - 32 x^4 - 8 x^6 + x^8$

One can check that our number is a root of this polynomial:

In[·]:= `FullSimplify[4096 - 512*x^2 - 32*x^4 -`
`        8*x^6 + x^8 /. x -> Sqrt[Sqrt[11] - 3] +`
`        Sqrt[Sqrt[11] + 5]]`
Out[·]:= `0`

We can also check that this polynomial cannot be factored over the rationals:

In[·]:= `Factor[4096 - 512*x^2 - 32*x^4 - 8*x^6 + x]`
Out[·]:= $4096 + x - 512 x^2 - 32 x^4 - 8 x^6$

Now let us try to do this problem without Mathematica®. Computing the minimal polynomial by hand is not easy in general. However, sometimes it is not needed. This is one of these cases. It turns out that it is sufficient to consider, along with the number $\sqrt{\sqrt{11} - 3} + \sqrt{\sqrt{11} + 5}$, its "conjugate", $\sqrt{\sqrt{11} - 3} - \sqrt{\sqrt{11} + 5}$, and compute the sum and the product of the two numbers.

In[·]:= `a = Sqrt[Sqrt[11] - 3] + Sqrt[Sqrt[11] + 5];`
`        b = Sqrt[Sqrt[11] - 3] - Sqrt[Sqrt[11] + 5];`

In[·]:= `a + b`
Out[·]:= $2\sqrt{-3 + \sqrt{11}}$

It is easy to show that $a + b$ is irrational and the product ab is rational:

In[·]:= `Expand[a*b]`
Out[·]:= `-8`

Clearly these two facts together imply that both numbers are irrational.
Note that this number is algebraic:

In[·]:= `Element[Sqrt[Sqrt[11] - 3] +`
`        Sqrt[Sqrt[11] + 5], Algebraics]`
Out[·]:= `True`

In[·]:= `MinimalPolynomial[Sqrt[Sqrt[11] - 3] +`
`        Sqrt[Sqrt[11] + 5], x]`
Out[·]:= $4096 - 512 x^2 - 32 x^4 - 8 x^6 + x^8$

In[·]:= `Clear[a, b]`

1.13 Non-algebraic equations

In this section we will discuss one more feature of Mathematica®, which appeared only in version 9, the ability to solve exactly non-algebraic equations, that is, equations in which functions other than polynomials and rational functions (and also expressions that can be reduced to polynomials, e. g., by a substitution) appear.

Consider, for example, the equation

In[·]:= Exp[x] == Sin[x] + x
Out[·]:= E^x == x + Sin[x]

This is, of course, a transcendental (non-algebraic) equation and most computer programs can only solve it approximately. If we try Solve or Reduce to solve this equation, without any additional information we will have

In[·]:= Solve[Exp[x] == Sin[x] + x, x]
  ··· Solve: This system cannot be solved with the methods available
   to Solve.
Out[·]:= Solve[E^x == x + Sin[x], x]

However, Mathematica® can prove that this equation has no real solutions:

In[·]:= Solve[Exp[x] == Sin[x] + x, x, Reals]
Out[·]:= {}

 Mathematica® can find a complex root, which we omit:

In[·]:= FindInstance[Exp[x] == Sin[x] + x, x];

It can also find exact complex solutions in any bounded region in $\mathbb{C}$. For instance, we have two solutions that lie in the unit disk:

In[·]:= sols = x /. Solve[Exp[x] == Sin[x] + x && Abs[x] <= 1,
  x]
Out[·]:= {Root[
  {-E$^{\#1}$ + Sin[#1] + #1 &, 0.6951202178045377839522875426609611653488601484275672699242 −
  0.7187235764741774958149153342820653950739435677373626299084 4 I}], Root[
  {-E$^{\#1}$ + Sin[#1] + #1 &, 0.6951202178045377839522875426609611653488601484275672699242 +
  0.7187235764741774958149153342820653950739435677373626299084 4 I}]}

Mathematica® uses its own system for naming and distinguishing these solutions. These solutions are represented by Root objects. In the case of algebraic equations Root objects are ordered and numbered. In the case of non-algebraic equations, instead of ordering and numbering them, Mathematica® gives an approximation which distinguishes different roots. These roots can of course be computed to an arbitrary precision:

In[·]:= N[First[sols], 3]
Out[·]:= 0.695 - 0.719 I

Traditional methods (e. g., Newton's method) can only solve such equations approximately and require starting values. The function FindRoot uses Newton's method (or a variant of the secant method) to compute numerical solution to the equation with a given initial condition (which can be complex):

In[·]:= FindRoot[Exp[x] == Sin[x] + x, {x, 1 + I}]
Out[·]:= {x -> 0.69512 + 0.718724 I}

Unlike Solve and Reduce, FindRoot can return a "false" root:

In[·]:= FindRoot[Exp[x] == Sin[x] + x, {x, 1}]
   ··· FindRoot: The line search decreased the step size to within
   tolerance specified by AccuracyGoal and PrecisionGoal but was
   unable to find a sufficient decrease in the merit function.
   You may need more than MachinePrecision digits of working
   precision to meet these tolerances.
Out[·]:= {x -> 0.601346}

Note that FindRoot can also work with systems of equations.

In general, Mathematica® can find the roots in a bounded interval of any equation in one variable given by an analytic function (i. e., one that has a convergent Taylor expansion at every point of its domain; see Chapter 6). The methods that it uses are too advanced to be described in this book, but we shall make use of this ability later on.

1.14 Sequences of real numbers and their limits

A *sequence* of real numbers $a_1, a_2, \ldots, a_n, \ldots$ is a function $a : \mathbb{N} \to \mathbb{R}$, where $a_n = a(n)$. We shall usually denote such a sequence by $\{a_n\}$. Mathematica® has no special objects corresponding to the notion of a sequence (the function Sequence is a programming rather than a mathematical concept) but sometimes it is useful to simulate the behavior of sequences by using lists. For example

In[·]:= Table[(-1)^n*(1/n), {n, 1, 8}]
Out[·]:= $\left\{-1, \frac{1}{2}, -\frac{1}{3}, \frac{1}{4}, -\frac{1}{5}, \frac{1}{6}, -\frac{1}{7}, \frac{1}{8}\right\}$

can be viewed as the beginning (the first eight terms) of the sequence $\{(-1)^n/n\}$.

A sequence $\{a_n\}$ is said to be *monotone* if it is increasing, i. e., $a_{n+1} \geq a_n$ for all n, or if it is decreasing, i. e., $a_{n+1} \leq a_n$ for all n. If a strict inequality holds, then we say that the sequence is *strictly monotone*.

We say that a sequence $\{a_n\}$ tends to a limit $a \in \mathbb{R}$, as n tends to infinity, and write $\lim_{n\to\infty} a_n = a$ (or $a_n \to a$), if given any $\varepsilon > 0$ we can find an integer $n_0(\varepsilon)$ (the notation is meant to remind us that this integer depends on ε) such that $|a_n - a| < \varepsilon$ for all $n \geq n_0(\varepsilon)$. We say that a sequence $\{a_n\}$ tends to ∞ ($-\infty$) if given any $M \in \mathbb{R}$ there exists $n_0(M)$ such that $a_n > M$ ($a_n < M$) for all $n \geq n_0(M)$. A sequence $\{a_n\}$ is called *convergent* if it has a limit in $\mathbb{R}$. When $\lim_{n\to\infty} a_n = \infty$ (or $-\infty$) we say that $\{a_n\}$ is divergent to ∞ or $-\infty$.

One can easily prove the following well-known properties of limits (see, for instance [14, Section 2], [13]).

Properties of limits:

(i)   The limit of a convergent sequence is unique, i. e., if $a_n \to a$ and $a_n \to b$, then $a = b$.

(ii) If $a_n \to a$ and $f : \mathbb{N} \to \mathbb{N}$ is such that $\lim_{n\to\infty} f(n) = \infty$, then $a_{f(n)} \to a$.

(iii) If $a_n \to a$ and $b_n \to b$, then $a_n + b_n \to a + b$.

(iv) If $a_n \to a$ and $b_n \to b$, then $a_n b_n \to ab$.

(v)   If $a_n \to a$ and $a_n \neq 0$ for each n and $a \neq 0$, then $a_n^{-1} \to a^{-1}$.

(vi) If $a_n \leq A$ for each n and $a_n \to a$, then $a \leq A$; if $a_n \geq B$ for each n and $a_n \to a$, then $a \geq B$.

(vii) If $a_n = c$ for all n, then $a_n \to c$.

Statements (i) to (vi) are also valid when a and b are either ∞ or $-\infty$, provided that the operations on the right hand side (involving a and b) are well-defined (i. e., Mathematica® does not return Indeterminate).

Note also the following important fact: a sequence $\{a_n\}_{n=1}^{\infty}$ has a limit a if and only if the sequence $\{a_n\}_{n=k}^{\infty}$ has a as its limit for some $k \geq 1$. In other words, when considering limits of sequences one can always ignore any finite number of initial terms of the sequence.

The following theorems are also very important (see, for instance, [14, Section 2] for their proofs).

Theorem 1 (The Monotone Convergence Theorem). *Every bounded monotonic sequence $\{a_n\}$ of real numbers converges. Equivalently, a monotonic sequence converges if and only if it is bounded.*

Moreover, if the sequence is increasing, then $\lim_{n\to\infty} a_n = \sup(\{a_n\})$. If it is decreasing, then $\lim_{n\to\infty} a_n = \inf(\{a_n\})$.

Theorem 2 (Bolzano–Weierstrass Theorem). *Every bounded sequence has a convergent subsequence.*

In versions of Mathematica® older than version 11.2, limits of sequences are computed using the function Limit, which is actually intended for computing limits of sequences that are obtained by restricting a continuous function $a : \mathbb{R} \to \mathbb{R}$ to $\mathbb{N}$. In version 11.2 of Mathematica® a new function DiscreteLimit was introduced, which can compute the limits of many sequences which do not arise in this way. For example

In[·]:= DiscreteLimit[1/Prime[n], n -> Infinity]
Out[·]:= 0

while

In[·]:= Limit[1/Prime[n], n -> Infinity]

Out[·]:= $\lim\limits_{n\to\infty} \dfrac{1}{\text{Prime}[n]}$

Note that when an expression is formatted in TraditionalForm the functions Discrete-Limit and Limit will look exactly the same:

In[·]:= TraditionalForm[DiscreteLimit[f[n], n -> Infinity]]

Out[·]://TraditionalForm= $\lim\limits_{n\to\infty} f(n)$

In[·]:= TraditionalForm[Limit[f[n], n -> Infinity]]

Out[·]://TraditionalForm= $\lim\limits_{n\to\infty} f(n)$

There also exist DiscreteMaxLimit (MaxLimit) and DiscreteMinLimit (MinLimit) for computing limit superior and limit inferior.

We will use DiscreteLimit although almost all examples that we will consider can equally well be solved with Limit. However, here is an example[6] where the results after using DiscreteLimit and Limit are different:

In[·]:= DiscreteLimit[n*Sin[2*Pi*E*n!], n -> Infinity]

Out[·]:= 2π

In[·]:= Limit[n*Sin[2*Pi*E*n!], n -> Infinity]

Out[·]:= Indeterminate

Here we assume that the reader is familiar with the number π and trigonometric functions.

Now let us try some other limits.

In[·]:= DiscreteLimit[(1 - 1/n)^n, n -> Infinity]

Out[·]:= $\frac{1}{E}$

In[·]:= DiscreteLimit[(1.0001 - 1/n)^n, n -> Infinity]

Out[·]:= ∞

In[·]:= DiscreteLimit[(0.99999 + 1/n)^n, n -> Infinity]

Out[·]:= 0

Note that Mathematica®11.3 returned an approximate zero (0.). This is a general principle: if a formula contains any *inexact number* (all decimal fractions represent inexact numbers in Mathematica®) the answer will always be inexact. Generally it is better to avoid using inexact numbers in symbolic computations. The following expression gives the exact answer to the example above:

In[·]:= DiscreteLimit[(99999/100000 + 1/n)^n, n -> Infinity]

Out[·]:= 0

---

6 https://math.stackexchange.com/questions/76097/what-is-the-limit-of-n-sin-2-pi-cdot-e-cdot-n-as-n-goes-to-infinity

Mathematica®'s lists can be viewed as finite parts of sequences. Operations that can be performed on sequences can also be performed on lists and it is often helpful actually see their effects. Mathematica® has a large number of functions for manipulating lists and they can be helpful for understanding sequences. We illustrate this with a few examples.

As mentioned before, we can view a part of a sequence by using Table:

In[·]:= Table[(1/8)*(9 + (-1)^(1 + n)), {n, 1, 6}]
Out[·]:= {5/4, 1, 5/4, 1, 5/4, 1}

We can use Simplify on the general term of a sequence:

In[·]:= Assuming[Mod[n, 2] == 0, Simplify[
         (1/8)*(9 + (-1)^(1 + n))]]
Out[·]:= 1

In[·]:= Assuming[Mod[n, 2] == 1, Simplify[
         (1/8)*(9 + (-1)^(1 + n))]]
Out[·]:= 5/4

We can construct sequences associated with a given sequence. Particularly important are the sequence of differences (the discrete analogue of the derivative of a function) and the sequence of "partial sums" (the discrete analogue of the integral):

In[·]:= Accumulate[{a, b, c, d, e}]
Out[·]:= {a, a + b, a + b + c, a + b + c + d, a + b + c + d + e}

which is equivalent to

In[·]:= FoldList[Plus, {a, b, c, d}]
Out[·]:= {a, a + b, a + b + c, a + b + c + d}

In[·]:= Differences[{a, b, c, d, e}]
Out[·]:= {-a + b, -b + c, -c + d, -d + e}

In[·]:= Ratios[{a, b, c, d, e}]
Out[·]:= {b/a, c/b, d/c, e/d}

In[·]:= Partition[{a, b, c, d, e}, 2]
Out[·]:= {{a, b}, {c, d}}

There are also a number of useful function which work with lists to construct new lists:

In[·]:= Remove[f]; Remove[x]

In[·]:= Comap[{f, g, h}, x]
Out[·]:= {f[x], g[x], h[x]}

In[·]:= ComapApply[{f, g, h}, {1, 2}]
Out[·]:= {f[1, 2], g[1, 2], h[1, 2]}

In[·]:= MapThread[f, {{a, b, c}, {x, y, z}}]
Out[·]:= {f[a, x], f[b, y], f[c, z]}

There is also a function that interlaces elements of two lists:

In[·]:= l1 = Table[n^2, {n, 1, 5}]
Out[·]:= {1, 4, 9, 16, 25}

In[·]:= l2 = Range[5]
Out[·]:= {1, 2, 3, 4, 5}

In[·]:= Riffle[l1, l2]
Out[·]:= {1, 1, 4, 2, 9, 3, 16, 4, 25, 5}

The function Thread is useful for creating a list for substitutions:

In[·]:= Thread[{a, b} -> {2, 2}]
Out[·]:= {a -> 2, b -> 2}

Elements in the list can be sorted using the function Sort, for example, the following list with numbers is sorted by absolute value:

In[·]:= Sort[{-11, 10, 2, 1, -4}, Abs[#1] < Abs[#2] &]
Out[·]:= {1, 2, -4, 10, -11}

The function Select picks out all elements satisfying certain condition:

In[·]:= Select[{1, 2, 4, 7, 6, 2}, EvenQ]
Out[·]:= {2, 4, 6, 2}

Another, more refined way to select elements is to use the functions Sow and Reap. In the following example we take the first 100 natural numbers and make two lists: one consisting of primes with remainder 1 modulo 4 and the other of primes with remainder 3. We use for them Sow with tag 1 or tag 3 depending on their remainder after division by 4 and then use Reap with the tag 1 to obtain the list of primes with remainder 1. Of course we could have used other tags, they only serve to distinguish the lists.

In[·]:= Reap[Block[{n = 2}, While[n < 100, Which[
          PrimeQ[n] && Mod[n, 4] == 1, Sow[n, 1],
          PrimeQ[n] && Mod[n, 4] == 3, Sow[n, 3]];
        n++]], 1][[2]]
Out[·]:= {{5, 13, 17, 29, 37, 41, 53, 61, 73, 89, 97}}

1.14.1 Example

Often a result is much easier to prove if we somehow discover the answer (by guessing, using a graphic program or Mathematica®). Let us compute $\lfloor nx \rfloor /n$, where x is a real number.

In[·]:= DiscreteLimit[Floor[n*x]/n, n -> Infinity]
Out[·]:= x

In the case of limits, it is often the case that to prove that $a_n \to x$ it is easier to prove the equivalent result $|a_n - x| \to 0$. In this case we have

$$0 \le \left| x - \frac{\lfloor nx \rfloor}{n} \right| = \frac{|nx - \lfloor nx \rfloor|}{n} \le \frac{1}{n}$$

since $|nx - \lfloor nx \rfloor| \le 1$.

1.14.2 Example: the number *e*

Let us start with a well-known limit

In[·]:= DiscreteLimit[(1 + 1/n)^n, n -> Infinity]
Out[·]:= E

This is usually taken as the definition of the number *e*. However, one needs to prove that the limit exists. The argument is based on applying the Monotone Convergence Theorem. To prove that the sequence $(1+1/n)^n$ is convergent it is enough to show that it is bounded and increasing. Let us first verify this with Mathematica®:

In[·]:= Reduce[(1 + 1/n)^n < 3 && n >= 1, Integers]
Out[·]:= n ∈ ℤ && n ≥ 1

This means that Mathematica® reduced the given condition to "*n* is an integer greater than or equal to 1". Now let us verify that the sequence is increasing. This will take longer:

In[·]:= Reduce[(1 + 1/n)^n < (1 + 1/(n + 1))^(n + 1) &&
          n >= 1, Integers]
Out[·]:= n ∈ ℤ && n ≥ 1

As we have explained in the preface we will (usually) not try to describe the algorithms that Mathematica® uses to obtain results such as these. This is a fascinating subject but it is not within the scope of this book.

Let us now consider the mathematical way of proving that the sequence is bounded, i. e., $(1 + 1/n)^n < 3$. For $n = 1$ the inequality is obvious. The proof for $n \ge 2$ is very simple and uses only the *Binomial Theorem*. This theorem is known to Mathematica® in the form

In[·]:= Sum[Binomial[n, i]*a^i*b^(n - i), {i, 0, n}]
Out[·]:= $(a + b)^n$

where the binomial coefficient $\binom{n}{k}$ is given by

```
In[·]:= FunctionExpand[Binomial[n, k], Assumptions ->
           {Element[{n, k}, Integers], n >= k >= 0}]
```
$$Out[\cdot]:= \frac{\text{Gamma}[1 + n]}{\text{Gamma}[1 + k]\,\text{Gamma}[1 - k + n]}$$

Mathematica® expresses the answer in terms of the Gamma function, which is the so called "special function" defined for all complex (and in particular real) numbers which has the property

```
In[·]:= FullSimplify[Gamma[n + 1], Assumptions ->
           Element[n, Integers] && n >= 0]
```
$Out[\cdot]:=$ n!

Now we can easily see that for $n \geq 2$

$$\left(1 + \frac{1}{n}\right)^n = 1 + 1 + \frac{1}{2!}\left(1 - \frac{1}{n}\right) + \frac{1}{3!}\left(1 - \frac{1}{n}\right)\left(1 - \frac{2}{n}\right) + \cdots + \frac{1}{n^n}$$
$$\leq 1 + 1 + \frac{1}{2!} + \frac{1}{3!} + \cdots \leq 1 + 1 + \frac{1}{2} + \frac{1}{2^2} + \cdots = 3.$$

The same binomial formula shows at once that

$$\left(1 + \frac{1}{n}\right)^n \leq \left(1 + \frac{1}{n+1}\right)^{n+1}$$

(just expanding both sides and comparing the terms). This kind of method cannot be used by Mathematica®. Although it can expand expressions such as

```
In[·]:= Expand[(a + b)^5]
```
$Out[\cdot]:= a^5 + 5a^4b + 10a^3b^2 + 10a^2b^3 + 5ab^4 + b^5$

it cannot do so with a general n:

```
In[·]:= Expand[(a + b)^n]
```
$Out[\cdot]:= (a + b)^n$

1.14.3 Example

Find the limit of the sequence

```
In[·]:= Limit[Sum[k^4 + 1/Sqrt[k], {k, 1, n}]^(1/n),
           n -> Infinity]
```
$Out[\cdot]:=$ 1

Note that if we slightly change the problem, then the following computations take more time:

In[·]:= `Limit[Sum[k^999 + 1/Sqrt[k], {k, 1, n}]^(1/n),`
     `n -> Infinity]`
Out[·]:= 1

The reason is that although for a human both problems look very similar and the idea of the proof is the same (we can apply the squeeze theorem in both cases, for instance, in the second case $1 \le n \le (nn^{1000})^{1/n} = n^{1001/n} = 1$), for Mathematica® the second problem is much more computationally difficult then the first one. Mathematica® first attempts to compute the sum (which is in fact not necessary as we have seen) and then takes the limit.

1.14.4 Example: the Euler–Mascheroni constant

Let

$$r_n = \left(\sum_{k=1}^{n} \frac{1}{k} \right) - \ln(n).$$

The limit $\lim_{n\to\infty} r_n$ is the Euler–Mascheroni constant.[7] Mathematica® knows that:

In[·]:= `r[n_] := Sum[1/k, {k, 1, n}] - Log[n]`

In[·]:= `Limit[r[n], n -> Infinity]`
Out[·]:= `EulerGamma`

Let us consider the proof that the sequence is convergent. We define one more sequence:

In[·]:= `s[n_] := Sum[1/k, {k, 1, n - 1}] - Log[n]`

We first show that s_n is an increasing and r_n decreasing sequence.

In[·]:= `s[n + 1] - s[n]`
Out[·]:= `-HarmonicNumber[-1 + n] + HarmonicNumber[n] +`
       `Log[n] - Log[1 + n]`

Mathematica® returns expression involving the function `HarmonicNumber`, which is inconvenient for our purpose. However, `FucitonExpand` will convert it into a more suitable form:

In[·]:= `FunctionExpand[%]`
Out[·]:= `1/n + Log[n] - Log[1 + n]`

This is equal to $1/n - \ln(1 + 1/n)$. Since $x \ge \ln(1 + x)$, this is non-negative. We could prove it directly using `Reduce`:

---

7 https://brilliant.org/wiki/euler-mascheroni-constant/

In[·]:= Reduce[1/n + Log[n] - Log[1 + n] >= 0, PositiveIntegers]
Out[·]:= n ∈ ℤ&&n ≥ 1

Similarly,

In[·]:= Reduce[FunctionExpand[r[n] - r[n + 1]] >= 0,
 n, PositiveIntegers]
Out[·]:= n ∈ ℤ&&n ≥ 1

Now let us compute the difference

In[·]:= FunctionExpand[r[n] - s[n]]
Out[·]:= 1/n

This shows that s_n is increasing and bounded above, hence convergent, while r_n is decreasing and bounded below, hence also convergent. As the difference between them tends to zero, they converge to the same limit.

1.15 Supremum and infimum

To state the last of the axioms that characterize the real numbers we need the concepts of the *supremum* (*least upper bound*) and the *infimum* (*greatest lower bound*) of a subset of ℝ.

Let A be a non-empty subset of ℝ. A real number M such that for all $a \in A$ we have $a \leq M$ is called an *upper bound* for A. Similarly, if there is a number m such that $a \geq m$ for all $a \in A$, we say that m is a *lower bound* of A. Of course a set may have no upper or lower bound (when it has one we say that it is bounded above or below) and if it has one, it has infinitely many bounds. If M is an upper bound of A and also for every upper bound N we have $N \geq M$, then we say that M is the *supremum* (or *the least upper bound*) of A. The infimum (greatest lower bound) is defined similarly. When the set A has only finitely many elements, it has both the maximum and the minimum element and these are precisely the supremum and the infimum of the set. Infinite subsets of ℝ do not, of course, necessarily have maximum or minimum elements (think of open intervals). However, we have the *Axiom of Completeness*: every non-empty subset $A \subset ℝ$ which is bounded above has a supremum. By replacing A with $-A = \{-a \mid a \in ℝ\}$ we see that the analogous statement holds also for sets bounded below, with supremum replaced by infimum. For sets that are not bounded above we consider the supremum to be equal to ∞ and for sets not bounded below we consider the infimum to be equal to $-\infty$. The supremum of the empty set is $-\infty$ and the infimum is ∞.

A reformulation of the definition of supremum (similarly for infimum) that is often useful in practical problems is: "a supremum of a set A is an upper bound M of A, such that $M - \varepsilon$ is not an upper bound of A for every $\varepsilon > 0$".

Let us consider the set

$$A = \{x \in \mathbb{R} \mid x^2 < 2\}. \tag{1.1}$$

Actually, we can check with Mathematica® that there exists an element satisfying our definition of the least upper bound, that is, one that is larger than all reals whose squares are less than two, but such that anything smaller than it no longer has this property:

In[·]:= Reduce[Exists[M, ForAll[y, y^2 <= 2, y <= M && ForAll[
          e, e > 0, Exists[z, z^2 <= 2 && M - e <= z]]]]]
Out[·]:= True

A typical example of the use of the axiom of completeness is proving the existence of roots, for example the square root of 2. It was already known to Ancient Greeks that there is no rational number whose square is equal to 2 (the discovery of this fact is attributed to Pythagoras). Now let us sketch an argument [12] which uses the axiom of completeness to show that there exists a real number whose square is 2. We consider the set A defined by (1.1). This set is bounded above. For example, 3 is an upper bound, since for any integer larger than 3 its square will be larger than 2. Hence the set has a least upper bound a. Now we need to show that $a^2 = 2$, which is done by indirect argument [12, Prop. 4.2]. Of course this information is built into Mathematica®:

In[·]:= Reduce[Exists[x, Element[x, Rationals], x^2 == 2]]
Out[·]:= False

In[·]:= Reduce[Exists[x, Element[x, Reals], x^2 == 2]]
Out[·]:= True

Finding suprema and infima is one of the main tasks of analysis. A large part of this book will be concerned with this problem and we will see that only in certain special cases complete solutions are possible.

Mathematica® has many built-in functions that try to find suprema and infima automatically. The most important ones are Maximize and Minimize. As their names suggest, these functions try to find the maximum and minimum of a set (normally the set of values of some function).

Suppose, for example, we want to find the supremum and the infimum of the set of real numbers of the form $x^3 - 3x^2 + 1$, where x is a real number in the open interval $(-1, 1)$. We can do this with Maximize and Minimize as follows:

In[·]:= Maximize[{x^3 - 3*x^2 + 1, -1 < x < 1}, x]
Out[·]:= {1, {x -> 0}}

The set (function) has a maximum value 1 attained at the point where $x = 0$.

In[·]:= Minimize[{x^3 - 3*x^2 + 1, -1 < x < 1}, x]

     ··· Minimize: Warning: there is no minimum in the region in

     which the objective function is defined and the constraints

     are satisfied; returning a result on the boundary.

Out[·]:= {-3, {x -> -1}}

This time Mathematica® produces a warning that the minimum is not attained at any point within the region but it returns a point on the boundary and a minimum is attained there. That, as we shall prove later, is exactly equivalent to the statement that the infimum is –3.

To verify the result, we can plot the graph of the function $x \mapsto x^3 - 3x^2 + 1$:

In[·]:= Plot[x^3 - 3*x^2 + 1, {x, -1, 1}]

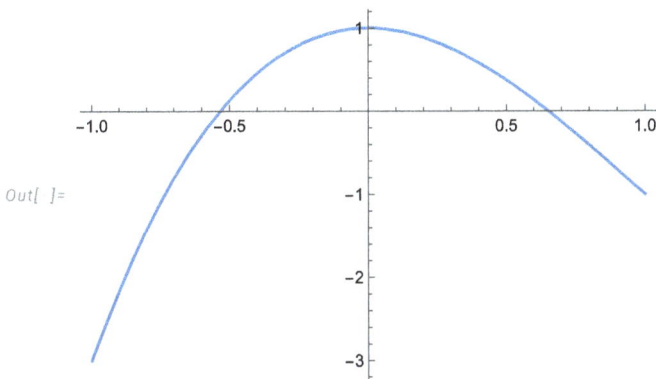

Out[ ]=

Figure 1.10

We can solve the same problem by using quantifiers. In the first case both the supremum and the infimum are actually attained, i. e., they are the maximum and minimum values. In the second case the infimum is not attained.

In[·]:= Reduce[Exists[x, -1 <= x <= 1, y == x^3 - 3*x^2 + 1],

     Reals]

Out[·]:= $-3 \le y \le 1$

In[·]:= Reduce[Exists[x, -1 < x < 1, y == x^3 - 3*x^2 + 1],

     Reals]

Out[·]:= $-3 < y \le 1$

Maximize and Minimize work with any number of variables. For example:

In[·]:= Maximize[{x + y, x^2 + y^2 == 1}, {x, y}]

Out[·]:= $\{\sqrt{2}, \{x \rightarrow \frac{1}{\sqrt{2}}, y \rightarrow \frac{1}{\sqrt{2}}\}$

Even if the function is real-valued but is expressed in terms of complex numbers (which quite often happens when using Mathematica®) Maximize will not work:

In[·]:= `Maximize[{Abs[2*x + I*y], x^2 + y^2== 1}, {x, y}]`
··· `Maximize: The objective function Abs[2 x+I y] contains a`
`nonreal constant I.`
Out[·]:= `Maximize[{Abs[2 x + I y], x^2 + y^2 == 1}, {x, y}]`

In such cases it is necessary to use the function `ComplexExpand` to explicitly express the function in terms of the real and imaginary parts of the complex variable:

In[·]:= `ComplexExpand[Abs[2*x + I*y]]`
Out[·]:= $\sqrt{4x^2 + y^2}$

In[·]:= `Maximize[{ComplexExpand[Abs[2*x + I*y]],`
`        x^2 + y^2 == 1}, {x, y}]`
Out[·]:= `{2, {x -> -1, y -> 0}}`

1.15.1 Example

Let us find the supremum and infimum of the following sets of real numbers:

$$A = \left\{ \frac{7n + 9k}{9n + 7k}, n > 0, k > 0 \right\}, \quad B = \left\{ \frac{7n + 9k}{9n + 7k}, n \in \mathbb{N}, k \in \mathbb{N} \right\}. \qquad (1.2)$$

We represent the elements of the sets A and B as the values of two functions of two variables: one defined on the set of pairs of natural numbers, i. e., the Cartesian product $\mathbb{N} \times \mathbb{N}$, the other defined on the Cartesian product $\mathbb{R}_+ \times \mathbb{R}_+$, the set of pairs of real numbers greater than 0. Our problem is to find the extrema of function values on these two sets.

The ability to compute limits can be very useful in proving that an upper (lower) bound of a set is a supremum (infimum). This is due to the following trivial lemma.

Lemma 3. *Let A be a set and M its upper bound. Then M is the supremum of A if and only if there exists a sequence* $\{a_n\}, a_n \in A$ *such that* $\lim_{n \to \infty} a_n = M$.

Let us see how this is applied in practice. We approach the problem first by looking at it graphically. We will use the functions `Plot3D` and `DiscretePlot3D`. First, let us look at the graphs which correspond to both sets:

In[·]:= `g1 = Plot3D[(7*n + 9*k)/(9*n + 7*k), {n, 1, 20},`
`        {k, 1, 20}, ColorFunction -> ({Opacity[0.5], Green}`
`        & ), Mesh -> False];`

In[·]:= `g2 = DiscretePlot3D[(7*n + 9*k)/(9*n + 7*k), {n, 1, 20},`
`        {k, 1, 20}, AxesLabel -> {"n", "k"}];`

In[·]:= `Show[g2, g1]`

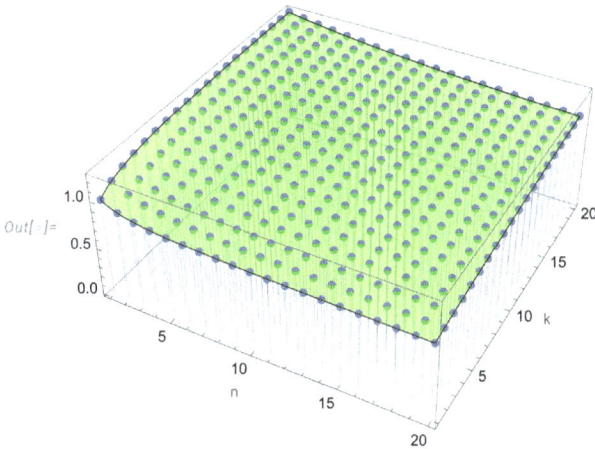

Figure 1.11

Clearly all the points belonging to the set B (represented by the discrete lattice of points) belong to A (represented by the surface). Looking at the graph and possibly expanding the range of n and k it seems that the surface and the lattice lie inside the box. The question is how to find the size of the enclosing box. In fact, for the surface this can be done using Maximize and Minimize:

In[·]:= Maximize[{(7*n + 9*k)/(9*n + 7*k), n > 0, k > 0},
          {n, k}]

   ··· Maximize::natt: The maximum is not attained at any point satisfying the

   given constraints.

Out[·]:= {$\frac{9}{7}$, {n -> Indeterminate, k -> ComplexInfinity}}

In[·]:= Minimize[{(7*n + 9*k)/(9*n + 7*k), n > 0, k > 0},
          {n, k}]

   ··· Minimize::natt: The minimum is not attained at any point satisfying the

   given constraints.

Out[·]:= {$\frac{7}{9}$, {n -> ComplexInfinity, k -> Indeterminate}}

Sometimes Mathematica® is not able to solve the problem over an unbounded domain of the function.

Clearly in our example 9/7 and 7/9 are the upper and lower bounds for the surface. Let us prove this by using Reduce:

In[·]:= Reduce[7/9 < (7*n + 9*k)/(9*n + 7*k) < 9/7 &&
          n > 0 && k > 0, {n, k}]

Out[·]:= n > 0 && k > 0

This says that the inequality "reduces" to the conditions $n > 0$ and $k > 0$, which mean that indeed these are an upper and lower bound (the reader can prove it easily by hand).

So once we have found upper and lower bounds, we can try to prove that the first is the least one (i. e., supremum) and the second is the greatest one (i. e., infimum). So now, according to the lemma, we need to find sequences lying in A and B whose limits are the two bounds. In fact, our sequences will be in B (and hence also in A) which will prove that the bounds are actually the supremum and infimum of both sets.

For the first sequence we fix $k = 1$, and we consider the sequence $a_n = (7n + 9)/(9n + 7)$. We have

In[·]:= `DiscreteLimit[(7*n + 9)/(9*n + 7), n -> Infinity]`

Out[·]:= $\dfrac{7}{9}$

Similarly we fix $n = 1$ and define $b_k = (7 + 9k)/(9 + 7k)$. We have

In[·]:= `DiscreteLimit[(7 + 9*k)/(9 + 7*k), k -> Infinity]`

Out[·]:= $\dfrac{9}{7}$

We can illustrate this as follows.

In[·]:= `g3 = Graphics3D[{Red, PointSize[0.025], Point[Table[`
`    {n, 1, (7*n + 9)/(9*n + 7)}, {n, 1, 20}]]}];`

In[·]:= `g4 = Graphics3D[{Purple, PointSize[0.025], Point[`
`    Table[{1, k, (7 + 9*k)/(9 + 7*k)}, {k, 1, 20}]]}];`

In[·]:= `Show[g2, g1, g3, g4]`

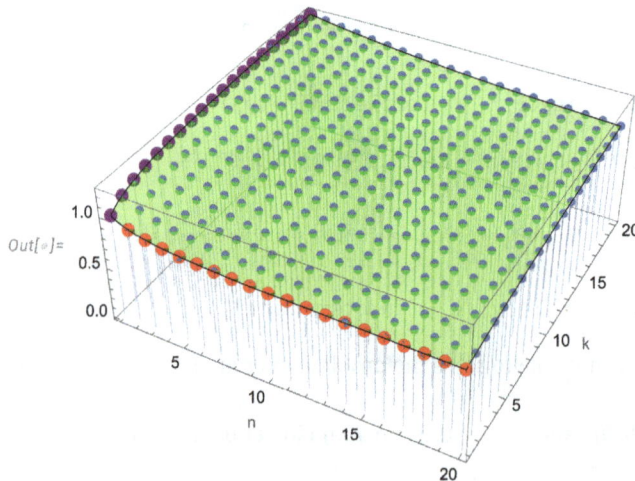

Out[·]=

Figure 1.12

Of course there are many other sequences of points in B whose limits are 9/7 and 7/9. Also, the supremum and infimum of the surface and a lattice of points need not to be the same, as in this example.

Note an important fact: although Mathematica® was able to solve the problem of finding the supremum and infimum of the set of points lying on the surface, it cannot do the same with the lattice (although it can solve the problem on any finite part of the lattice). Indeed, the expression

In[·]:= Maximize[{(7*n + 9*k)/(9*n + 7*k), Element[{n, k},
　　　　 Integers] && n > 0 && k > 0}, {n, k}]

is returned not evaluated.

Although Maximize (and Minimize) do work over the integers in finite cases,

In[·]:= Maximize[{(9*k + 7*n)/(7*k + 9*n), Element[{n, k},
　　　　 Integers] && 10 > n > 0 && 10 > k > 0}, {n, k}]
Out[·]:= {$\frac{97}{79}$, {n -> 1, k -> 10}}

the problem of finding the least and greatest bounds of infinite discrete sets of points is actually harder. The function NMaximize, which uses numerical approximate methods, is more successful, but its answer is not guaranteed to be correct even approximately:

In[·]:= NMaximize[{(7*n + 9*k)/(9*n + 7*k), Element[{n, k},
　　　　 Integers] && n > 0 && k > 0}, {n, k}]
　　　 ··· NMaximize::ubndv: One or more elements of the solution vector is unbounded
　　　　 but the objective function approaches a bounded optimal value.
Out[·]:= {1.28571, {n -> Indeterminate, k -> ComplexInfinity}}

Here only the first five digits are correct, as compared with

In[·]:= N[9/7]
Out[·]:= 1.28571

1.15.2 Example

Let us similarly find the supremum and infimum of the following set

$$A = \left\{ \frac{n - m}{n + m}, \, n, m \in \mathbb{N} \right\}$$

over integers. First it is easy to show that the set is bounded from above and below:

In[·]:= Reduce[-1 <= (n - m)/(n + m) <= 1,
　　　　 {n, m}, PositiveIntegers]
Out[·]:= (n|m) ∈ ℤ&&n ≥ 1&&m ≥ 1

We can also find sequences from this set that converge to those numbers:

In[·]:= Limit[(n - m)/(n + m) /. m -> n^2, n -> Infinity]
Out[·]:= -1

In[·]:= Limit[(n - m)/(n + m) /. m -> 1, n -> Infinity]
Out[·]:= 1

Alternatively, we can show that $1 - \varepsilon$ cannot be the upper bound for any $\varepsilon > 0$. We look for some n, m such that

$$1 - \varepsilon < \frac{n - m}{n + m}.$$

We can put $m = 1$ and find n such that

$$1 - \varepsilon < \frac{n - 1}{n + 1}.$$

In[·]:= Reduce[1 - e < (n - 1)/(n + 1) &&
          e > 0 && n > 1, n]
Out[·]:= $(0 < e \le 1 \&\& n > \frac{2-e}{e}) \,||\, (e > 1 \&\& n > 1)$

As we have already seen, Mathematica® can find the extrema for the continuous case:

In[·]:= Maximize[{(x - y)/(x + y), x >= 1,
          y >= 1}, {x, y}]
Out[·]:= {1, {x -> ComplexInfinity, y -> Indeterminate}}

In[·]:= Minimize[{(x - y)/(x + y), x >= 1,
          y >= 1}, {x, y}]
Out[·]:= {-1, {x -> Indeterminate, y -> ComplexInfinity}}

For discrete case Mathematica® can do it only for finite sets:

In[·]:= Maximize[{(x - y)/(x + y), 200 >= x, 200 >= y,
          Element[{x, y}, PositiveIntegers]}, {x, y}]
Out[·]:= {199/201, {x -> 200, y -> 1}}

In[·]:= Maximize[{(x - y)/(x + y), Element[
          {x, y}, PositiveIntegers]}, {x, y}]
Out[·]:= Maximize[{(x - y)/(x + y), Element[x |
          y, Integers] && x > 0 && y > 0}, {x, y}]

We can try to use the function NMaximize, which uses numerical methods:

In[·]:= NMaximize[{(x - y)/(x + y), Element[{x, y},
          PositiveIntegers]}, {x, y}]
          ··· NMaximize::ubndv: One or more elements of the solution vector is
          unbounded but the objective function approaches a bounded
          optimal value.
Out[·]:= {1., {x -> ComplexInfinity, y -> Indeterminate}}

Note that the result of NMaximize does not guarantee that a global extremum has been found.

1.15.3 Example

Find the extrema of the following set

$$A = \left\{ \frac{(n+m)^2}{2^{nm}}, \ n, m \in \mathbb{N} \right\}.$$

For finite n and m we can compute

In[·]:= Maximize[{(n + m)^2/2^(n*m), n <= 10,
    m <= 10}, {n, m}, PositiveIntegers]
Out[·]:= {9/4, {n -> 1, m -> 2}}

In[·]:= N[Minimize[{(n + m)^2/2^(n*m), n <= 10,
    m <= 10}, {n, m}, PositiveIntegers]]
Out[·]:= {3.155443620884047*^-28, {n -> 10., m -> 10.}}

In[·]:= Show[{Plot3D[(x + y)^2/2^{x*y}, {x, 1, 10},
    {y, 1, 10}, ColorFunction -> Function[
    {x, y, z}, {Opacity[0.2], Red}],
    PlotRange -> All], DiscretePlot3D[
    (x + y)^2/2^{x*y}, {x, 1, 10}, {y, 1, 10},
    PlotRange -> All, PlotStyle -> Green],
    Graphics3D[{Blue, PointSize[0.03], Point[
    {1, 2, 9/4}], Orange, Point[{10, 10,
    3.15*^-28}]}]}]

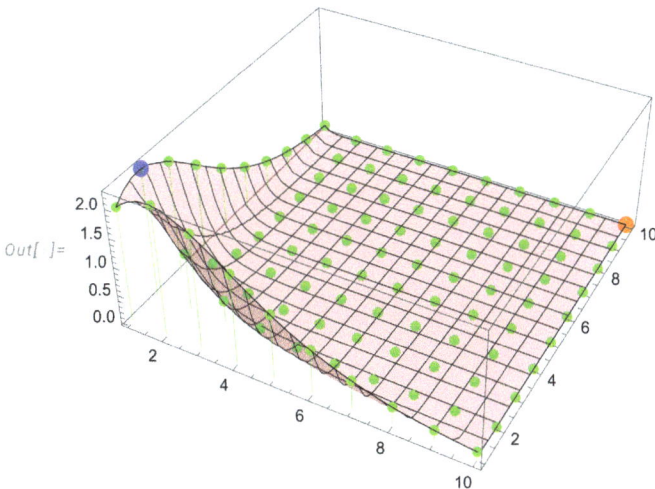

Out[ ]=

Figure 1.13

It is obvious that the infimum is zero for large n or m:

In[·]:= Limit[(n + 1)^2/2^n, n -> Infinity]
Out[·]:= 0

Let us show that supremum is equal to 9/4. We can prove first that

In[·]:= Reduce[2^n > (n + 1)^2, n, PositiveIntegers]
Out[·]:= n ∈ ℤ&&n ≥ 6

This can also be proved by induction. Then $2^{nm} > (nm+1)^2 \geq (n+m)^2$, since $nm+1 > n+m$ is equivalent to $(n-1)m > n-1$. Hence,

$$\frac{(n + m)^2}{2^{nm}} < 1$$

for $nm > 6$. Hence, we need to check values of (n, m) which satisfy $nm < 6$.

In[·]:= ls = Cases[Flatten[Table[{n, m}, {n, 1, 6},
        {m, 1, 6}], 1], {x_, y_} /; x*y < 6];

In[·]:= rules = ({n -> #1[[1]], m -> #1[[2]]} &) /@ ls;

In[·]:= vals = (n + m)^2/2^(n*m) /. rules;

In[·]:= Max[%]
Out[·]:= 9/4

2 Recursive sequences, discrete dynamical systems and their limits

A *recursive sequence* is the sequence which is defined as follows: several initial values are given and the remaining ones are defined in terms of the previous ones. In this chapter we will show how to define recursive sequences by various methods, solve recurrence equations and also digress to other related topics (computing with inexact numbers with a given precision and so on).

Let us start with a simple case and define a recurrence

$$a_1 = 1, \quad a_n = a_{n-1} + 1.$$

Let us first look at some explicit terms of this sequence. There are a number of ways to do this. The fastest is the built-in function RecurrenceTable:

```
In[·]:= RecurrenceTable[{a[n] == a[n - 1] + 1, a[1] == 1},
          a, {n, 1, 5}]
Out[·]:= {1, 2, 3, 4, 5}
```

To get a better feeling for what is actually going on when Mathematica® builds this kind of lists, we will first do so using more basic programming constructs. Readers familiar with other programming languages might be tempted to use looping constructs. It is possible to do so with Mathematica® because Mathematica® has looping constructs such as Do, For and While, but it is generally better, both from the point of view of elegance and efficiency, to try to write programs in functional style and avoid them.

There are other ways to define recurrences. For instance a direct recursive definition:

```
In[·]:= Clear[a]
In[·]:= a[1] := 1; a[n_] := a[n - 1] + 1
In[·]:= a[3]
Out[·]:= 3
```

Mathematica® remembers only the following information about a:

```
In[·]:= ?a
Out[·]:= Global`a
          a[1] := 1
          a[n_] := a[n-1] + 1
```

and, therefore, computations might become slow.

We can modify the definition of $a[n]$ above by using *dynamic programming* [16] such that in each evaluation the value of $a[n]$ will be memorized and therefore the calculations become faster at the expense of using a little memory.

https://doi.org/10.1515/9783111533063-002

```
In[·]:= Clear[a]
In[·]:= a[1] := 1; a[n_] := a[n] = a[n - 1] + 1
In[·]:= a[3]
Out[·]:= 3
```

Now Mathematica® remembers the following data about *a*:

```
In[·]:= ?a
Out[·]:= Global`a
          a[1] := 1
          a[2] = 2
          a[2] = 3
          a[n_] := a[n] = a[n-1] + 1
```

Mathematica® has several functions designed to perform iterations. Here we will use Nest and NestList. The function Nest takes three arguments and works as follows:

```
In[·]:= Clear[f]
In[·]:= Nest[f, a, 5]
Out[·]:= f[f[f[f[f[a]]]]]
```

There is also a related function, NestList, which returns a list of all the values (beginning with the initial one) obtained during iteration:

```
In[·]:= NestList[f, a, 4]
Out[·]:= {a, f[a], f[f[a]], f[f[f[a]]], f[f[f[f[a]]]]}
```

We will see later on how to apply these functions to recursive sequences. One more application of the function Nest from Mathematica®'s documentation is a bifurcation diagram for an iterated logistic map:

```
In[·]:= ListPlot[Table[Thread[{r, Nest[r*#1*(1 - #1) & ,
         Range[0, 1, 0.01], 100]}], {r, 0, 4, 0.01}]]
```

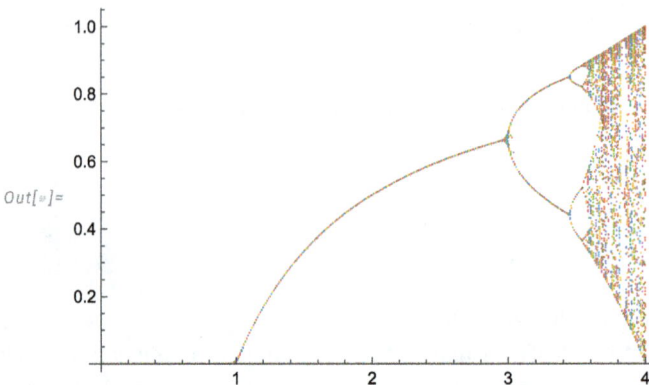

Figure 2.1

Here

In[·]:= Nest[r*#1*(1 - #1) & , Range[0, 1, 0.45], 2]
Out[·]:= {0., 0.2475 (1 - 0.2475 r) r^2, 0.09 (1 - 0.09 r) r^2}

In[·]:= Thread[{r, Nest[r*#1*(1 - #1) & , Range[0, 1, 0.4], 2]}]
Out[·]:= {{r, 0.}, {r, 0.24 (1 - 0.24 r) r^2}, {r, 0.16 (1 - 0.16 r) r^2}}

See also "Applying Functions Repeatedly".[1]

Mathematica® has a function RSolve designed to solve recursive equations. Only certain types of equations can be solved as most are too difficult for RSolve:

In[·]:= Clear[a]; RSolve[{a[n + 1] == a[n]*(a[n] + 1), a[1] == 1},
        a[n], n]
Out[·]:= RSolve[{a[n + 1] == a[n] (a[n] + 1), a[1] == 1},
        a[n], n]

Fortunately, usually one can find the limits of sequences defined by such equations (when they exist) without solving the equations.

Finally, we would like to note that the global variable $IterationLimit gives the maximum length of evaluation chain used in trying to evaluate any expression. The default value is

In[·]:= $IterationLimit
Out[·]:= 4096

We can temporarily change it using the function Block:

In[·]:= f[0] = 1; f[p_] := f[p - 1]
In[·]:= Block[{$IterationLimit = 21}, f[20]]
        ⋯ $IterationLimit: Iteration limit of 21 exceeded.
Out[·]:= TerminatedEvaluation[IterationLimit]

With a higher $IterationLimit, the function can be computed:

In[·]:= Block[{$IterationLimit = 22}, f[20]]
Out[·]:= 1

The iteration limit protects against infinite loops:

In[·]:= f[x_] := f[x + 1]
In[·]:= f[1]
        ⋯ $IterationLimit: Iteration limit of 4096 exceeded.
Out[·]:= TerminatedEvaluation[IterationLimit]

One can even set $IterationLimit equal to Infinity, however, there is a risk of getting into an infinite loop in computations in more complicated examples as a result of a programming error:

---

1 https://reference.wolfram.com/language/tutorial/FunctionalOperations.html#23850

In[·]:= f[0] = 1; f[n_] := f[n - 1]
In[·]:= Block[{$IterationLimit = Infinity}, f[10^5]]
Out[·]:= 1

$RecursionLimit is a similar global variable. It gives the current limit on the number of levels of recursion that the Wolfram Language can use. It limits infinite recursion with circular definitions:

In[·]:= x = x + 1
··· $RecursionLimit: Recursion depth of 1024 exceeded.
Out[·]:= 1 + TerminatedEvaluation[RecursionLimit]

and can also be temporarily reset using the function Block, as we shall see later. $RecursionLimit has by default the value:

In[·]:= $RecursionLimit
Out[·]:= 1024

The reason for this is to stop infinite recursion from occurring as a result of programming errors. However, this can sometimes be inconvenient, as we can see from the following example.

In[·]:= Clear[Fib]; Fib[1] = 1; Fib[2] = 1; Fib[n_] :=
        Fib[n] = Fib[n - 1] + Fib[n - 2];

In[·]:= Fib[3000]
··· $RecursionLimit: Recursion depth of 1024 exceeded.
··· $RecursionLimit: Recursion depth of 1024 exceeded.
Out[·]:= 2 TerminatedEvaluation[RecursionLimit]

The value of 1024 for $RecursionLimit prevents the code from working. Using Block we can temporarily change this value:

In[·]:= Clear[Fib]; Fib[1] = 1; Fib[2] = 1; Fib[n_] :=
        Fib[n] = Fib[n - 1] + Fib[n - 2];

In[·]:= Block[{$RecursionLimit = 10000}, Fib[3000]];

We omitted the large numerical output. Note that the global value of $RecursionLimit remains unchanged:

In[·]:= $RecursionLimit
Out[·]:= 1024

In the next examples we will consider more complicated recurrences and methods of computing the terms of recursive sequences and finding their limits.

2.1 Example

Consider the sequence $\{a_n\}$ given by

$$a_1 = 1, \quad a_{n+1} = \frac{1}{2}\left(a_n + \frac{2}{a_n}\right). \tag{2.3}$$

Using RecurrenceTable we can find the first few terms of the sequence:

In[·]:= RecurrenceTable[{a[n] == (1/2)*(a[n - 1] + 2/a[n - 1]),
    a[1] == 1}, a, {n, 1, 5}]

Out[·]:= $\left\{1, \dfrac{3}{2}, \dfrac{17}{12}, \dfrac{577}{408}, \dfrac{665857}{470832}\right\}$

Let us compare how much time is required to compute a few terms of the sequence by using a simple recursive definition approach and by using the so called dynamic programming technique. We set the initial value

In[·]:= a[1] = 1
Out[·]:= 1

and then we define the following elements of the sequence by

In[·]:= a[n_] := (1/2)*(a[n - 1] + 2/a[n - 1])

To compute several elements of the sequence and to check the amount of time Mathematica® needs for this computation we use the function Timing (the function Timing evaluates the expression and returns a list of the time in seconds used and the result obtained):

In[·]:= Timing[Table[a[i], {i, 1, 20}];]
Out[·]:= {2.48438, Null}

In the expression above the semi-column actually suppresses the result of the evaluation (since it is very long) and therefore we see Null as the second element of the output. We see that it takes a lot of time to compute the elements of the sequence. Using dynamic programming we compute the same result faster:

In[·]:= Clear[a]
In[·]:= a[1] = 1; a[n_] := a[n] = (1/2)*(a[n - 1] + 2/a[n - 1])
In[·]:= Timing[Table[a[i], {i, 1, 20}];]
Out[·]:= {0.03125, Null}

The terms of our sequence (2.3) are obtained by iterating the function Function[x,1/2 (x+2/x)] and evaluating it at 1. In an abbreviated (pure) form the function can be written as 1/2(# + 2/#)&.

In[·]:= NestList[(1/2)*(#1 + 2/#1) & , 1, 5]

Out[·]:= $\left\{1, \dfrac{3}{2}, \dfrac{17}{12}, \dfrac{577}{408}, \dfrac{665857}{470832}, \dfrac{886731088897}{627013566048}\right\}$

Because we started with an exact initial value 1 and our function also contains only exact numbers, the whole computation is performed using exact arithmetic, which is very slow and not very informative. Let us replace the initial value 1 by an *inexact number* (1.). All computations will then be with inexact numbers. Recall that in Mathematica® decimals always represent inexact numbers, or more precisely so called `MachinePrecision` numbers.

In[·]:= `NestList[(1/2)*(#1 + 2/#1) & , 1., 6]`
Out[·]:= `{1., 1.5, 1.41667, 1.41422, 1.41421, 1.41421, 1.41421}`

It now looks like the iteration stabilizes after only a few steps. We can see better what is going on by expressing the output in `InputForm`, because otherwise Mathematica® (by default) shows only a few leading digits:

In[·]:= `InputForm[NestList[(1/2)*(#1 + 2/#1) & , 1., 6]]`
Out[·]//InputForm= `{1., 1.5, 1.4166666666666665, 1.4142156862745097,`
　　　　`1.4142135623746899, 1.414213562373095, 1.414213562373095}`

The sequence still stabilizes, but that is only because a point is reached where the difference between successive numbers is smaller than the precision with which the computations have been performed (the so called machine precision, which is in this case 15 digits).

In Mathematica® `Precision` of an approximate number is not exactly equal to the number of digits after the decimal point and in fact it is not even an integer but a real number which measures the relative error in the number. It is approximately equal to the number of significant digits.

When the differences between successive terms becomes sufficiently small Mathematica® no longer distinguishes between them and the sequence appears to reach a fixed point of iteration. We could reach this fixed point quicker if instead of `Nest` (`NestList`) we use the function `FixedPoint` (`FixedPointList`) which performs iteration until a fixed point is reached. `FixedPoint` starts with an expression and then applies the function repeatedly until the result no longer changes:

In[·]:= `InputForm[FixedPoint[(1/2)*(#1 + 2/#1) &, 1.]]`
Out[·]//InputForm= `1.414213562373095`

In[·]:= `InputForm[FixedPointList[(1/2)*(#1 + 2/#1) & , 1.]]`
Out[·]//InputForm= `{1., 1.5, 1.4166666666666665, 1.4142156862745097,`
　　　　`1.4142135623746899, 1.414213562373095, 1.414213562373095}`

In[·]:= `Length[%]`
Out[·]:= `7`

It took six iterations to reach the fixed point. Of course we have to be careful not to use `FixedPoint` with an exact initial value since the exact sequence has no fixed point! (We could, however, use the form `FixedPoint[a,f,n]` which will stop either after a fixed point is reached or after *n* iterations, whichever happens sooner.)

In addition to MachinePrecision arithmetic Mathematica® also has *arbitrary precision arithmetic.* A number *a* with precision *p* is most easily entered as N[a, p]. Here *a* should either be an exact number (which has precision infinity) or a number with precision higher than *p*. For example,

In[·]:= N[1, 5]
Out[·]:= 1.0000

In[·]:= N[1/4, 20]
Out[·]:= 0.25000000000000000000

In[·]:= N[%, 5]
Out[·]:= 0.25000

In[·]:= Precision[%]
Out[·]:= 5.

Note the difference between using $MachinePrecision and MachinePrecision:

In[·]:= N[1, $MachinePrecision]
Out[·]:= 1.000000000000000

In[·]:= Precision[%]
Out[·]:= 15.9546

But

In[·]:= N[1, MachinePrecision]
Out[·]:= 1.

In[·]:= Precision[%]
Out[·]:= MachinePrecision

As we have already mentioned before, exact numbers have infinite precision:

In[·]:= Precision[1]
Out[·]:= ∞

We can compute our example with higher precision:

In[·]:= FixedPointList[(1/2)*(#1 + 2/#1) & , N[1, 30]]
Out[·]:= {1.00000000000000000000000000000,
           1.50000000000000000000000000000,
           1.41666666666666666666666666667,
           1.41421568627450980392156862745,
           1.41421356237468991062629557889,
           1.41421356237309504880168962350,
           1.41421356237309504880168872421,
           1.41421356237309504880168872421}

In[·]:= Length[%]
Out[·]:= 8

In fact, we need 100 digits of precision to make the sequence longer by just a few elements:

In[·]:= Length[FixedPointList[(1/2)*(#1 + 2/#1) & , N[1, 100]]]
Out[·]:= 10

Mathematica® can completely solve our recurrence:

In[·]:= Clear[a]; sol[n_] = First[a[n] /. RSolve[{a[n] == (1/2)*
        (a[n - 1] + 2/a[n - 1]), a[1] == 1}, a[n], n]]
··· Solve: Inverse functions are being used by Solve, so some
solutions may not be found; use Reduce for complete
solution information.

Out[·]:= $\sqrt{2}\,\mathrm{Coth}\left[2^{-1+n}\,\mathrm{ArcCoth}\left[\dfrac{1}{\sqrt{2}}\right]\right]$

The function RSolve warns us that there may be also other solutions; we shall ignore this warning and compute the limit.

In[·]:= DiscreteLimit[sol[n], n -> Infinity]
Out[·]:= $\sqrt{2}$

Without any initial conditions we can obtain the general solution

In[·]:= First[a[n] /. RSolve[{a[n] == (1/2)*
        (a[n - 1] + 2/a[n - 1])}, a[n], n]]
Out[·]:= $-\mathrm{I}\,\sqrt{2}\,\mathrm{Cot}[2^n\,C[1]]$

Here C[1] is an arbitrary constant to be determined from initial conditions. However, this answer is not satisfactory to us, since we get the complex i in the answer and we do not clearly see the dependence on initial data. To make things clearer, we obtain the answer which explicitly depends on the initial conditions:

In[·]:= First[a[n] /. RSolve[{a[n] == (1/2)*
        (a[n - 1] + 2/a[n - 1]), a[1] == x}, a[n], n]]
··· Solve: Inverse functions are being used by Solve, so some
solutions may not be found; use Reduce for complete
solution information.

Out[·]:= $\sqrt{2}\,\mathrm{Coth}\left[2^{-1+n}\,\mathrm{ArcCoth}\left[\dfrac{x}{\sqrt{2}}\right]\right]$

and we can check the limits:

In[·]:= DiscreteLimit[%, n -> Infinity, Assumptions -> {x > 0}]
Out[·]:= $\sqrt{2}$

In[·]:= DiscreteLimit[%%, n -> Infinity, Assumptions -> {x < 0}]
Out[·]:= $-\sqrt{2}$

Note that the last two evaluations take a lot of time.

Now we are going to find the limit of our sequence (2.3) without solving the recurrence equation. We will start by assuming that the sequence a_n has a limit. Under this assumption, we will find all the possible values that this limit can have. Finally we will prove this assumption (that the limit exists) and determine its actual values.

Consider the sequence given by the equation $a_{n+1} = 1/2(a_n + 2/a_n)$ and assume it has a limit a (possibly ∞ or $-\infty$). We can suppose that $a_n \neq 0$ for all n; otherwise the sequence could not be defined. We can consider the equation as an equality of two sequences $\{a_{n+1}\}$ and $\{1/2(a_n + 2/a_n)\}$. The sequence $\{a_{n+1}\}$ is a subsequence of $\{a_n\}$ and therefore has the same limit a. Looking at the equation we immediately see that $a \neq 0$. Then the sequence $\{1/2(a_n+2/a_n)\}$ converges to $1/2(a+2/a)$. Thus we obtain the equation

$$a = \frac{1}{2}\left(a + \frac{2}{a}\right).$$

Solving it we get:

In[·]:= a /. Solve[a == (1/2)*(a + 2/a), a, Reals]
Out[·]:= $\{-\sqrt{2}, \sqrt{2}\}$

There are also two other possibilities that we will still need to consider: ∞ and $-\infty$ (which also satisfy the equation).

We now turn to the next stage: proving the existence of a limit and finally determining its value. Now we will need the information (which we have not used so far) about the initial condition. We can prove the following statements. If $a_1 > 0$, then $a_n \geq \sqrt{2}$ for $n > 1$, which follows from the well-known inequality between the geometric and arithmetic means of two numbers: $1/2(a_n + 2/a_n) \geq \sqrt{2}$ and induction. If $a_1 < 0$, then $a_n \leq -\sqrt{2}$ for $n > 1$. The sequence is decreasing exactly when

In[·]:= Reduce[(1/2)*(x + 2/x) <= x, x]
Out[·]:= $-\sqrt{2} \leq x < 0 \lor x \geq \sqrt{2}$

Since $a_1 = 1$, we have $a_n \geq \sqrt{2}$ for $n > 1$ and the next term of the sequence a_{n+1} will be smaller and it can never fall below $\sqrt{2}$. Thus the sequence is monotone decreasing and bounded below by $\sqrt{2}$, hence it has a limit greater than or equal to $\sqrt{2}$. But the only possibilities are ∞ and $\sqrt{2}$ and since the sequence is decreasing, the only possibility for the limit is $\sqrt{2}$. Arguing in the same way, we see that if the initial value a_1 is negative, then the limit of the sequence is $-\sqrt{2}$.

It is natural to think of a sequence given by a recursive equation as of the motion of a particle, which at the starting time is at a_1, then moves to a_2, etc. The equation $a_{n+1} = f(a_n)$ (in our case $f(x) = 1/2(x + 2/x)$) can be thought of as an equation of motion of a particle, giving its position at time $n + 1$ in terms of its position at time n. This is known as a *discrete dynamical system*. A *fixed point* of a function $f : X \to X$, where $X \subset \mathbb{R}$, is a point $x \in X$ such that $f(x) = x$. In other words, if the particle is located at a fixed point it stops moving. As we will see later, when f is a continuous function, a limit of such a sequence will always have to be a fixed point of f. Of course a fixed point is always the limit of the constant sequence starting at that point.

Mathematica®'s ability to quickly create interactive visualizations using Manipulate (or Dynamic) makes it easy to study discrete dynamical systems even in cases when they are difficult to deal with mathematically. In the interactive illustration below we show the movement of a particle (red point) whose position is a_{n+1} after n iterations. Initially the slider for the number of iterations should be set to zero. The starting point can be chosen by moving the gray circle (locator in Mathematica®) to any point in the interval $[-4, 4]$ on the real line. By default the starting point is chosen as 2. The blue points correspond to the fixed points. Because it is difficult to manually move the initial point to the position of the fixed points, one can do it by clicking the plus sign in the upper right corner and choose one of the bookmarked positions. All motions of the particle are on the real line but for visual convenience we use 2-dimensional graphics. We also use the function Quiet to suppress any unwanted messages from Mathematica®. Once locator is used to choose the starting point, then by increasing the number of iterations we can watch the point move to a limit, which is one of the two fixed points: $\sqrt{2}$ and $-\sqrt{2}$.

```
In[·]:= Manipulate[Quiet[Module[{fix}, fix =
           FixedPoint[(1/2)*(#1 + 2/#1) & , First[p], m];
           Graphics[{PointSize[0.02], Point[p], Red,
           PointSize[0.02], Point[{fix, 0}], Blue, Point[
           {{Sqrt[2], 0}, {-Sqrt[2], 0}}]}], PlotRange ->
           {{-4, 4}, {-1, 1}}, Axes -> True]]], {{p, {2, 0}},
           {-4, 0}, {4, 0}, Locator}, {{m, 0, "number of
           iterations"}, 0, 10, 1, Appearance -> "Labeled"},
           Bookmarks -> {"fixed point 1" :> (p = {Sqrt[2], 0}),
           "fixed point 2" :> (p = {-Sqrt[2], 0})}]
```

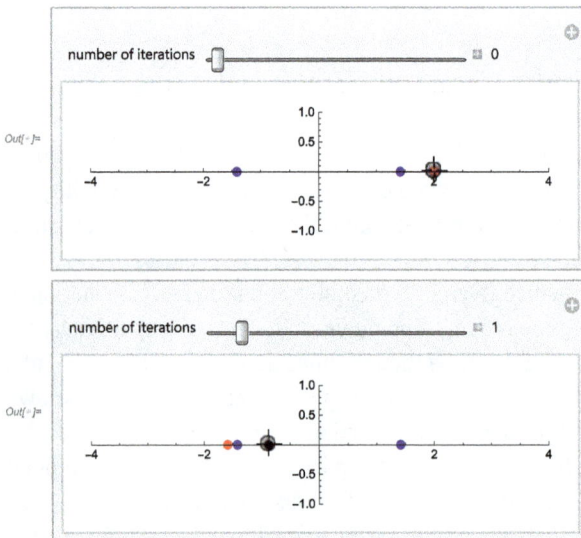

Figure 2.2

In the example above the sequence obtained from the recurrence eventually becomes monotonic, making it possible to use the Monotone Convergence Theorem to conclude the existence of a limit. We will now consider a case where this is not true.

2.2 Example: the Fibonacci sequence

The famous Fibonacci sequence is given by

$$a_1 = 1, \quad a_2 = 1, \quad a_{n+1} = a_n + a_{n-1}.$$

We can generate its terms with

In[·]:= RecurrenceTable[{a[n + 1] == a[n] + a[n - 1], a[1] == 1,
    a[2] == 1}, a, {n, 1, 14}]
Out[·]:= {1, 1, 2, 3, 5, 8, 13, 21, 34, 55, 89, 144, 233, 377}

We can also solve the recurrence with RSolve:

In[·]:= RSolve[{a[1] == 1, a[2] == 1, a[n] == a[n - 1] +
    a[n - 2]}, a[n], n]
Out[·]:= {{a[n] -> Fibonacci[n]}}

In[·]:= First[a[n] /. RSolve[{a[1] == 1, a[2] == 1,
    a[n] == a[n - 1] + a[n - 2]}, a[n], n]]
Out[·]:= Fibonacci[n]

Mathematica® returns a built-in symbol Fibonacci[n]. Below we use notation F_n for our sequence. We can expand it using

In[·]:= FunctionExpand[Fibonacci[n]]
Out[·]:= $\dfrac{\left(\frac{1}{2}\left(1 + \sqrt{5}\right)\right)^n - \left(\frac{2}{1+\sqrt{5}}\right)^n \cos[n\pi]}{\sqrt{5}}$

Note that this can be simplified under the assumption that n is either even or odd:

In[·]:= Simplify[FunctionExpand[Fibonacci[n]],
    Assumptions -> Element[n/2, Integers]]
Out[·]:= $\dfrac{-\left(\frac{2}{1+\sqrt{5}}\right)^n + \left(\frac{1}{2}\left(1 + \sqrt{5}\right)\right)^n}{\sqrt{5}}$

In[·]:= Simplify[FunctionExpand[Fibonacci[n]],
    Assumptions -> Element[(n + 1)/2, Integers]]
Out[·]:= $\dfrac{\left(\frac{2}{1+\sqrt{5}}\right)^n + \left(\frac{1}{2}\left(1 + \sqrt{5}\right)\right)^n}{\sqrt{5}}$

The sequence clearly diverges monotonically to ∞:

In[·]:= DiscreteLimit[Fibonacci[n], n -> Infinity]

Out[·]:= ∞

But the sequence

$$s_n = \frac{F_{n+1}}{F_n}$$

has an interesting limit:

In[·]:= DiscreteLimit[Fibonacci[n + 1]/Fibonacci[n],
          n -> Infinity]

Out[·]:= $\frac{1}{2}(1 + \sqrt{5})$

The sequence s_n is obviously given by the recurrence equation

$$s_1 = 1, \quad s_n = 1 + \frac{1}{s_{n-1}}.$$

It can be easily solved by RSolve

In[·]:= First[s[n] /. RSolve[{s[n]*s[n - 1] == s[n - 1] + 1,
          s[1] == 1}, s[n], n]]

Out[·]:= (Fibonacci[n] + Fibonacci[1 + n] - LucasL[n]
          - LucasL[1 + n])/(Fibonacci[1 + n] - LucasL[1 + n])

The answer involves another built-in function LucasL:

In[·]:= FunctionExpand[LucasL[n]]

Out[·]:= $\left(\frac{1}{2}(1 + \sqrt{5})\right)^n + \left(\frac{2}{1 + \sqrt{5}}\right)^n \text{Cos}[n\pi]$

Of course we can generate the terms of the sequence $\{s_n\}$ using RecurenceTable or FixedPointList:

In[·]:= N[RecurrenceTable[{s[n]*s[n - 1] == s[n - 1] + 1,
          s[1] == 1}, s, {n, 1, 8}]]

Out[·]:= {1., 2., 1.5, 1.66667, 1.6, 1.625, 1.61538, 1.61905}

In[·]:= InputForm[FixedPointList[1 + 1/#1 & , 1.]]

Out[·]://InputForm=

```
{1., 2., 1.5, 1.6666666666666665, 1.6, 1.625, 1.6153846153846154, 1.619047619047619,
 1.6176470588235294, 1.6181818181818182, 1.6179775280898876, 1.6180555555555556,
 1.6180257510729614, 1.6180371352785146, 1.6180327868852458, 1.618034447821682,
 1.618033813400125, 1.6180340557275543, 1.6180339631667064, 1.6180339985218035,
 1.618033985017358, 1.6180339901755971, 1.6180339882053252, 1.6180339889579018,
 1.6180339886704433, 1.6180339887802426, 1.618033988738303, 1.6180339887543225,
 1.6180339887482038, 1.6180339887505406, 1.6180339887496482, 1.618033988749989,
 1.618033988749859, 1.6180339887499087, 1.6180339887498896, 1.618033988749897,
 1.618033988749894, 1.6180339887498951, 1.618033988749895}
```

In[·]:= Length[%]
Out[·]:= 39

This sequence has reached a fixed point after 38 steps. We know that the exact sequence will never reach its limit, but if we take any approximation up to some precision p, eventually all the elements of the sequence will have the same first p digits. So the numbers will be equal to each other up to this precision p and Mathematica® will decide that it has reached the fixed point. But we know that this is not a real fixed point. Suppose we want to get a longer sequence by using more precision. The above computation is done using the so called machine arithmetic, which is done using MachinePrecision, whose value depends on the type of CPU used. As mentioned earlier, Mathematica® can also have arbitrary precision arithmetic. Recall (see the previous example in Section 2.1) that the easiest way to enter a number with a given precision p is N[a, p], where a is an exact number. For example the number

In[·]:= N[1, 20]
Out[·]:= 1.0000000000000000000

is 1 with 20 correct digits. If all numbers in a formula have non-machine precision, Mathematica® performs computations using its special model of arithmetic, which actually keeps track of the precision. Note however that the precision of an answer can either be higher or lower than the precision of the input. In fact, this is what happens in our example, the Precision of the result of each iteration is actually higher than that of the previous one:

In[·]:= rt = RecurrenceTable[{s[n + 1] == 1 + 1/s[n],
          s[1] == N[1, 1]}, s, {n, 1, 9}]
Out[·]:= {1., 2., 1.5, 1.7, 1.60, 1.63, 1.615, 1.619, 1.6176}

In[·]:= Map[Precision, rt]
Out[·]:= {1., 1.30103, 1.77815, 2.17609, 2.60206, 3.01703,
          3.43616, 3.8537, 4.27184}

We see that the precision is actually getting larger, hence a fixed point will never be reached! Hence, the next computation will never finish:

In[·]:= Length[FixedPointList[1 + 1/#1 & , N[1, 1]]]
Out[·]:= $Aborted

We can stop the computation above after inserting the third argument of the function FixedPointList:

In[·]:= Length[FixedPointList[1 + 1/#1 & , N[1, 1], 500]]
Out[·]:= 501

However, we can make sure that a fixed point is reached by forcing Mathematica® to perform all iterations using fixed precision (rather than its own variable precision arithmetic). The idea is to make the global variables $MinPrecision and $MaxPrecision equal to the same number, which will be our chosen fixed precision. By default these variables have the values

In[·]:= {$MinPrecision, $MaxPrecision}
Out[·]:= {0, ∞}

We can set them to other values; however, since we want to change them only temporarily, we use the construction Block, which is often used for such a purpose:

In[·]:= Length[Block[{$MinPrecision = 20, $MaxPrecision = 20},
        FixedPointList[1 + 1/#1 & , N[1, 20]]]]
Out[·]:= 49

We can, of course, easily calculate the possible "candidates" for the limit of the sequence

In[·]:= s /. Solve[s == 1 + 1/s, s, Reals]
Out[·]:= $\{\frac{1}{2}(1 - \sqrt{5}), \frac{1}{2}(1 + \sqrt{5})\}$

Note that neither ∞ nor $-\infty$ satisfies the equation. However proving the existence of a limit is this time harder (see below).

As in the previous example we can study how the sequence behaves by using Manipulate. Although the sequence does appear to converge to $(1 + \sqrt{5})/2$ for all starting points except for $(1 - \sqrt{5})/2$ (which is the other fixed point), the sequence does not appear to become monotone. Let us confirm this in our case with $s_1 = 1$:

In[·]:= Sign[Differences[RecurrenceTable[{s[n + 1] ==
        1 + 1/s[n], s[1] == 1}, s, {n, 1, 10}]]]
Out[·]:= {1, -1, 1, -1, 1, -1, 1, -1, 1}

The function Differences applied to a list gives the list of differences between successive elements. The function Sign returns the sign of a number, i. e., +1 for a positive number, −1 for a negative number and 0 for 0. So we can see that the signs of the differences between successive elements keep changing and the sequence is not monotonic. This suggests the following idea to prove the convergence.

Let us consider two subsequences of $\{s_n\}$, the subsequence of elements with an even index $\{s_{2n}\}$ and the subsequence of elements with an odd index $\{s_{2n+1}\}$. We will prove that both of these subsequences converge to the same limit. Then we will use the following simple lemma.

Lemma 4. *Let $\{a_n\}$ be a sequence of real numbers and suppose its subsequences $\{a_{2n}\}$ and $\{a_{2n+1}\}$ both have the same limit g. Then g is the limit of $\{a_n\}$.*

So to prove that $\{s_{2n}\}$ converges and that $\{s_{2n+1}\}$ converges, we will prove that $\{s_{2n+1}\}$ is increasing and $\{s_{2n}\}$ is decreasing. In other words, we want to show that $s_{2n+1} - s_{2n-1} > 0$ and $s_{2n+2} - s_{2n} < 0$. We proceed by induction on n. We can check directly that the result is true for $n = 1$:

In[·]:= RecurrenceTable[{s[n + 1] == 1 + 1/s[n], s[1] == 1},
      s, {n, 1, 4}]

Out[·]:= {1, 2, $\frac{3}{2}$, $\frac{5}{3}$}

So suppose now that it is true for $n = k$. We have

$$s_{2k+3} - s_{2k+1} = \frac{1}{s_{2k+2}} - \frac{1}{s_{2k}} > 0, \quad s_{2k+4} - s_{2k+2} = \frac{1}{s_{2k+3}} - \frac{1}{s_{2k+1}} < 0.$$

Hence both sequences are monotonic by induction. Note also that s_n is always bounded, since $1 \leq s_n \leq 2$ by induction (with the assumption that $s_1 = 1$). Hence both s_{2n} and s_{2n+1} have limits, say, a and b. The equation $s_{2n+1} = 1 + 1/s_{2n}$ implies that $b = 1 + 1/a$. The equation $s_{2n+2} = 1 + 1/s_{2n+1}$ implies $a = 1 + 1/b$. Solving these together gives

In[·]:= {a, b} /. Solve[{b == 1 + 1/a, a == 1 + 1/b}, {a, b}]

Out[·]:= {{$\frac{1}{2}(1 - \sqrt{5})$, $\frac{1}{2}(1 - \sqrt{5})$}, {$\frac{1}{2}(1 + \sqrt{5})$, $\frac{1}{2}(1 + \sqrt{5})$}}

Hence $a = b$ and the sequence $\{s_n\}$ is convergent.

2.3 Example

Find the limit of the sequence

$$\sqrt{2}, \ \sqrt{2 + \sqrt{2}}, \ \sqrt{2 + \sqrt{2 + \sqrt{2}}}, \dots .$$

We can write the elements of the sequence using the function Nest:

In[·]:= Nest[Sqrt[2 + #1] & , 0, 4]

Out[·]:= Sqrt[2 + Sqrt[2 + Sqrt[2 + Sqrt[2]]]]

In[·]:= N[%]

Out[·]:= 1.99037

or using the recurrence and dynamic programming

In[·]:= a[1] := 0

In[·]:= a[n_] := a[n] = Sqrt[a[n - 1] + 2]

In[·]:= a[4]

Out[·]:= Sqrt[2 + Sqrt[2 + Sqrt[2]]]

In[·]:= N[a[20]]

Out[·]:= 2.

The sequence is increasing and bounded. This can be proved by induction, since $a_1 = 0 \leq 2$ and assuming that $a_n \leq 2$ we have $a_{n+1} = \sqrt{a_n + 2} \leq \sqrt{2 + 2} = 2$. Hence the limit exists and is equal to a, which can be found from

```
In[·]:= a /. Solve[a == Sqrt[a + 2], a][[1]]
Out[·]:= 2
```

since sequences a_n and a_{n-1} have the same limit.

2.4 Example: the secant method

The secant method is a variant of Newton's method[2] to solve an equation $f(x) = 0$ for $f :$ $\mathbb{R} \to \mathbb{R}$. As the following example shows, one needs to choose carefully initial conditions for non-linear recurrence relations.

```
In[·]:= f[x_] := x^2 - 2
In[·]:= Clear[x]; x[0] = 2; x[1] = 2.2;

In[·]:= x[n_] := x[n] = x[n - 1] - f[x[n - 1]]*
          ((x[n - 1] - x[n - 2])/(f[x[n - 1]] - f[x[n - 2]]))

In[·]:= x[10]
Out[·]:= 1.41421

In[·]:= N[2^(1/2)]
Out[·]:= 1.41421
```

However,

```
In[·]:= x[11]
          ··· Power: Infinite expression 1/0. encountered.
          ··· Infinity: Indeterminate expression 0. ComplexInfinity encountered.
Out[·]:= Indeterminate
```

We could try to compute more terms, working with Mathematica®'s arbitrary precision arithmetic. However, giving our input even very high initial precision will not significantly improve the result.

```
In[·]:= Clear[x]; f[x_] := x^2 - 2;
In[·]:= x[0] = 2; x[1] = N[22/10, 1000];

In[·]:= x[n_] := x[n] = x[n - 1] - f[x[n - 1]]*
          ((x[n - 1] - x[n - 2])/(f[x[n - 1]] - f[x[n - 2]]))

In[·]:= N[x[17], 20]
Out[·]:= 1.4142135623730950488

In[·]:= Timing[Precision[x[17]]]
Out[·]:= {0., 994.162}
```

---

2 https://en.wikipedia.org/wiki/Secant_method

In[·]:= Timing[x[18]]

       ··· Power: Infinite expression 1/(0. * 10^{-994}) encountered.

       ··· Infinity: Indeterminate expression 0. * 10^{-994} ComplexInfinity encountered.

Out[·]:= {0., Indeterminate}

We can check that the precision of calculations actually decreases:

In[·]:= (Precision[#1] &) /@ {x[1], x[10], x[17]}

Out[·]:= {1000., 996.269, 994.162}

but this is not the cause of the problem. The problem is that the numbers become too small for any precision, whether variable or fixed. Using fixed precision with 10000 digits instead of 1000 lets us compute only a few more terms of the sequence:

In[·]:= Clear[x]; x[0] = 2; x[1] = N[22/10, 10000];

In[·]:= x[n_] := x[n] = x[n - 1] - f[x[n - 1]]*

      ((x[n - 1] - x[n - 2])/(f[x[n - 1]] - f[x[n - 2]]))

In[·]:= Block[{$MaxPrecision = 10000, $MinPrecision = 10000}, x[22]];

    We again omitted the large output above. However, the computation of x[23] is again not successful and Mathematica® returns Indeterminate.

    In general, a better approach with Mathematica® is to use exact input and then the function N with the desired precision to calculate the next elements of the sequence. In this way we can get more than 30 terms without difficulty:

In[·]:= Clear[x]; x[0] = 2; x[1] = 22/10;

In[·]:= x[n_] := x[n] = x[n - 1] - f[x[n - 1]]*

      ((x[n - 1] - x[n - 2])/(f[x[n - 1]] - f[x[n - 2]]))

In[·]:= Timing[N[x[18], 10]]

Out[·]:= {0., 1.414213562}

In[·]:= Timing[N[x[30], 10]]

Out[·]:= {3.64063, 1.414213562}

However, for $n > 30$ computations become very slow.

In[·]:= Timing[N[x[32], 10]]

Out[·]:= {8.57813, 1.414213562}

    In Mathematica® there are functions that perform iterations or recursion and they have options that control both maximum (and sometimes minimum) number of iterations or depth of recursion. One such example is FindRoot.

In[·]:= x /. FindRoot[x^2 - 2, {x, 2}]

Out[·]:= 1.41421

The option MaxIterations specifies the maximum number of iterations that should be tried in various built-in functions and algorithms. For instance,

```
In[·]:= x /. FindRoot[Exp[1/(x^2 - 1)], {x, 0.1}]
       ···FindRoot: Failed to converge to the requested accuracy or precision
          within 100 iterations.
Out[·]:= 0.995327
```

```
In[·]:= x /. FindRoot[Exp[1/(x^2 - 1)], {x, 0.1},
          MaxIterations -> 4000, WorkingPrecision -> 10]
Out[·]:= 0.9968826622
```

```
In[·]:= Clear[x]; x //. Sequence[x -> x + 1, MaxIterations -> 10]
       ···ReplaceRepeated: Exiting after x scanned 10 times.
Out[·]:= 10 + x
```

```
In[·]:= FindMinimum[x^3, x, MaxIterations -> 10]
       ···FindMinimum: Failed to converge to the requested accuracy or
          precision within 10 iterations.
Out[·]:= {3.348979766803838*^-7, {x -> 0.006944444444444442}}
```

```
In[·]:= FindMinimum[x^3, x, MaxIterations -> 1000]
Out[·]:= {1.6744185165715833*^-23, {x -> 2.5583188611367297*^-8}}
```

The function NIntegrate, which we shall discuss later in the book, has both options MaxRecursion and MinRecursion:

```
In[·]:= NIntegrate[1/Sqrt[Sin[x]], {x, 0, 10}]
Out[·]:= 10.3032 - 6.74255 I
```

```
In[·]:= NIntegrate[1/Sqrt[Sin[x]], {x, 0, 10}, MaxRecursion -> 1000]
Out[·]:= 10.4882 - 6.76947 I
```

With both outputs Mathematica® returned the following warning: "NIntegrate: Numerical integration converging too slowly; suspect one of the following: singularity, value of the integration is 0, highly oscillatory integrand, or WorkingPrecision too small". Moreover, in the first case Mathematica® returned the following additional warning: "NIntegrate: NIntegrate failed to converge to prescribed accuracy after 9 recursive bisections in x near x = 9.41547357990858'15.954589770191005. NIntegrate obtained 10.30316607463091' -6.742554183083223' I and 0.31689817716051966' for the integral and error estimates".

NIntegrate may miss sharp peaks of integrands:

```
In[·]:= Plot[Exp[-1000*x^2], {x, -1, 1}, PlotRange -> All]
       ···General: Exp[-999.918] is too small to represent as a normalized
          machine number; precision may be lost.
```

Out[ ]=

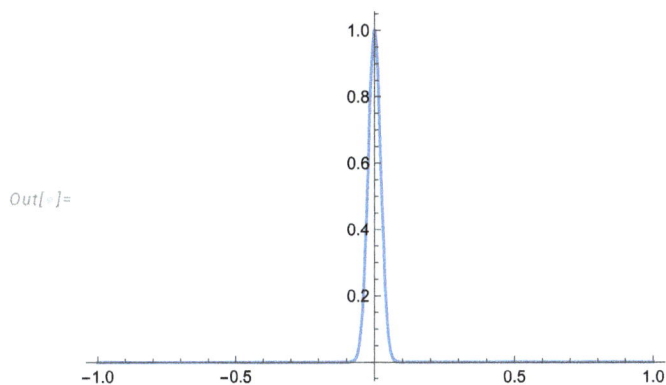

Figure 2.3

In[·]:= NIntegrate[Exp[-1000*x^2], {x, -50, 60}]

    ··· NIntegrate: Integral and error estimates are 0 on all

       integration subregions. Try increasing the value of the MinRecursion

       option. If value of integral may be 0, specify a finite value for the

       AccuracyGoal option.

Out[·]:= 0.

    Increasing MinRecursion forces a finer subdivision of the integration region:

In[·]:= NIntegrate[Exp[-1000*x^2], {x, -50, 60}, MinRecursion -> 9]

Out[·]:= 0.0560499

3 Series

In this chapter we define the concept of series and discuss their convergence. We consider various convergence tests and give several examples. We also discuss a number of related questions, for instance, Riemann's theorem on conditionally convergent series, divergent series and power series.

3.1 Sequences and series

We will model sequences by lists which can be thought of as sequences that are constant after a certain number of terms. Since Mathematica® cannot deal with symbolic lengths, we will fix n as some small integer but it will be clear that everything we say will work with an arbitrary n.

```
In[·]:= n = 4;
In[·]:= cc = Table[c[i], {i, 1, n}]
Out[·]:= {c[1], c[2], c[3], c[4]}
```

Mathematica® has two useful functions (in some sense almost inverse to each other) that take lists as arguments and which naturally extend to sequences. The first one is the function Differences, which has already appeared in Chapter 1 and which returns the list of differences between successive elements:

```
In[·]:= dd = Differences[cc]
Out[·]:= {-c[1] + c[2], -c[2] + c[3], -c[3] + c[4]}
```

A sequence convergent to 0 is called a *null sequence*. We have the following trivial but useful lemma.

Lemma 5. *If a sequence a_n is convergent, then its difference sequence is a null sequence.*

Results about limits of general sequences can often be reduced to results about null sequences, which often can be proved more easily than for general sequences. Here is a very useful fact about null sequences, which follows directly from the definition of limits.

Lemma 6. *A sequence a_n is a null sequence if and only if the sequence $|a_n|$ is a null sequence.*

The second useful function Accumulate is almost inverse to the function Differences:

```
In[·]:= Accumulate[cc]
Out[·]:= {c[1], c[1] + c[2], c[1] + c[2] + c[3],
          c[1] + c[2] + c[3] + c[4]}
```

https://doi.org/10.1515/9783111533063-003

For a sequence it gives a sequence of partial sums. Note that

In[·]:= Differences[Accumulate[cc]]
Out[·]:= {c[2], c[3], c[4]}

differs from the original sequence only be the first element; in other words, it is the original list without the first element:

In[·]:= Rest[cc]
Out[·]:= {c[2], c[3], c[4]}

If we perform the operations in the reverse order, we get

In[·]:= Accumulate[Differences[cc]]
Out[·]:= {-c[1] + c[2], -c[1] + c[3], -c[1] + c[4]}

This does not quite get us back to where we started but the following does:

In[·]:= Accumulate[Differences[cc]] + c[1]
Out[·]:= {c[2], c[3], c[4]}

or

In[·]:= Accumulate[Prepend[Differences[cc], c[1]]]
Out[·]:= {c[1], c[2], c[3], c[4]}

Since removing or adding a finite number of elements of a sequence does not make any difference to the limit, we can think of the functions Differences and Accumulate as essentially inverse to each other.

Accumulate is equivalent to a particular case of FoldList, but is optimized for numerical computation:

In[·]:= FoldList[Plus, {a, b, c, d}]
Out[·]:= {a, a + b, a + b + c, a + b + c + d}

In[·]:= Accumulate[{a, b, c, d}]
Out[·]:= {a, a + b, a + b + c, a + b + c + d}

In[·]:= ls = RandomReal[{1, 10}, 10^7];

In[·]:= Timing[Accumulate[ls];]
Out[·]:= {0.015625, Null}

In[·]:= Timing[FoldList[Plus, ls];]
Out[·]:= {0.53125, Null}

Note that the head does not need to be List:

In[·]:= Accumulate[f[a, b, c, d]]
Out[·]:= f[a, a + b, a + b + c, a + b + c + d]

A series is often informally thought of as an infinite sum, but there are several risks in this approach and we do not recommend it. Instead, we define the notion of a series as in [7]. By a *series* we mean the sequence of partial sums of some sequence. If this sequence of partial sums has a limit, it is called the *sum of the series*. If the limit is a real number we say that the series is *convergent* to that number. Therefore, we clearly distinguish between two objects: the series and its sum. The latter may or may not exist. Unfortunately, it is customary to denote both the series and its sum by the same symbol $\sum_{n=1}^{\infty} a_n$ and we have to be careful in interpreting the notation.

The partial sums of a series $\sum_{n=0}^{\infty} a_n$ are simply the solutions of the recurrence equation $s_1 = a_0, s_{n+1} = s_n + a_n$. Thus, instead of the function Accumulate, we can use the function RecurrenceTable:

```
In[·]:= Clear[n]
In[·]:= RecurrenceTable[{s[1] == a[0], s[n + 1] ==
            s[n] + a[n]}, s, {n, 1, 3}]
Out[·]:= {a[0], a[0] + a[1], a[0] + a[1] + a[2]}
```

It follows from Lemma 5 that a necessary condition for a series associated with a_n to converge is that a_n is a null sequence. However, we will soon see that the series of a null sequence is not always convergent (e. g., the harmonic series).

3.2 The functions Sum and NSum

Mathematica®'s function Sum has multiple uses. First of all, it can be used to simply add up a list of numbers, e. g.,

```
In[·]:= Sum[1, {10}]
Out[·]:= 10
```

The index in Sum can run over an arbitrary list:

```
In[·]:= Sum[i^2, {i, {2, 4, 7, 9, 11}}]
Out[·]:= 271
```

The function Sum is able to compute many finite sums explicitly, sometimes expressing them in terms of special functions. For example,

```
In[·]:= Sum[i^3, {i, 1, n}]
```
$Out[·]:= \dfrac{1}{4}n^2(1 + n)^2$

```
In[·]:= Sum[i, {i, 1, n, 2}]
```
$Out[·]:= \left(1 + \text{Floor}\left[\dfrac{1}{2}(-1 + n)\right]\right)^2$

```
In[·]:= Sum[1/k^2, {k, 1, n}]
Out[·]:= HarmonicNumber[n, 2]
```

This means that the sequence of partial sums of a sequence a_i can also be computed as

In[·]:= Table[Sum[a[i], {i, 1, k}], {k, 1, 3}]
Out[·]:= {a[1], a[1] + a[2], a[1] + a[2] + a[3]}

The sum of a series can be computed in Mathematica® simply by combining the functions Sum and Limit:

In[·]:= DiscreteLimit[Sum[1/k^2, {k, 1, n}], n -> Infinity]
Out[·]:= $\dfrac{\pi^2}{6}$

In fact, one can obtain the result more simply:

In[·]:= Sum[1/k^2, {k, 1, Infinity}]
Out[·]:= $\dfrac{\pi^2}{6}$

Note, however, that while in the case of convergent series there is no difference between the outputs above in Mathematica®, in the case of divergent series there is one. For example,

In[·]:= Sum[1/n, {n, 1, Infinity}]
··· Sum: Sum does not converge.
Out[·]:= $\displaystyle\sum_{n=1}^{\infty} \dfrac{1}{n}$

In[·]:= DiscreteLimit[Sum[1/n, {n, 1, k}], k -> Infinity]
Out[·]:= ∞

In[·]:= Sum[(-1)^n, {n, 1, Infinity}]
··· Sum: Sum does not converge.
Out[·]:= $\displaystyle\sum_{n=1}^{\infty} (-1)^n$

In[·]:= DiscreteLimit[Sum[(-1)^n, {n, 1, k}], k -> Infinity]
Out[·]:= Indeterminate

So we see that in these cases using an explicit DiscreteLimit gives us more information since it distinguishes between divergence to infinity and the other kind of divergence (non-existence of a limit).

Sometimes we want to truncate a series. We can do this by using TruncateSum:

In[·]:= ss = Sum[f[i], {i, 1, Infinity}];

In[·]:= TruncateSum[ss, 5]
Out[·]:= f[1] + f[2] + f[3] + f[4] + f[5]

This gives the same result as

In[·]:= Sum[f[i], {i, 1, 5}]
Out[·]:= f[1] + f[2] + f[3] + f[4] + f[5]

but with it one can save a lot of typing.

In[·]:= ss1 = Inactivate[Sum[1/(2*n), {n, 1, Infinity}] -
      Sum[1/(2*n + 1), {n, 0, Infinity}]];

In[·]:= TruncateSum[ss1, n]
Out[·]:= HarmonicNumber[n]/2 + (1/2)*(PolyGamma[0, 1/2]
      - PolyGamma[0, 1/2 + n])

In[·]:= N[TruncateSum[ss1, 50]]
Out[·]:= -0.688172

The function Sum has an option VerifyConvergence which by default is set to True. It will therefore always try to verify convergence and will inform us if it can show that the series is not convergent, as in the example above.

Now let us look at the sum

In[·]:= Sum[1/n^n, {n, 1, Infinity}]
Out[·]:= $\sum\limits_{n=1}^{\infty} n^{-n}$

Mathematica® returns the original input unchanged. What can we conclude from this? We can conclude that one of two things happened. One possibility is that Mathematica® verified that the series is convergent but could not find any closed form expression for the sum. The other possibility is that Mathematica® could not decide whether the series is convergent or not. To distinguish between these possibilities we need to use the function SumConvergence, which will be discussed in greater detail in Section 3.6.

In[·]:= SumConvergence[1/n^n, n]
Out[·]:= True

Now we know that the series is convergent but Mathematica® cannot find any closed form formula. In such cases we can now use the function NSum to compute the numerical value of this sum to any desired precision:

In[·]:= NSum[1/n^n, {n, 1, Infinity}, WorkingPrecision -> 20]
Out[·]:= 1.2912859970626635404

Alternatively we can write

In[·]:= N[Sum[1/n^n, {n, 1, Infinity}], 20]
Out[·]:= 1.2912859970626635404

Here the function Sum passed the task of computation to NSum. For simple problems we can use the second approach, but for more complicated ones some of the options offered by the function NSum may be needed.

Of course there many series for which Mathematica® cannot decide whether they are convergent or not. For instance, in Section 3.5.6 we shall prove that the next series is convergent but Mathematica® cannot do it:

In[·]:= Sum[Sin[1/n]*Sin[n], {n, 1, Infinity}]

Out[·]:= $\sum\limits_{n=1}^{\infty}$ Sin $\left[\dfrac{1}{n}\right]$ Sin[n]

Again the function Sum returns the input. But this time Mathematica® cannot decide whether this series converges:

In[·]:= SumConvergence[Sin[1/n]*Sin[n], n]

Out[·]:= SumConvergence $\left[$Sin $\left[\dfrac{1}{n}\right]$ Sin[n], n $\right]$

If nevertheless we try to compute this sum numerically to a high degree of precision, i. e., we write N[Sum[Sin[1/n]*Sin[n], {n, 1, Infinity}], 30], we will get a lot of warning messages from Mathematica®. This is because Mathematica® fails to notice that the series has non-constant signs and is using a wrong computational method. By changing the method, we can successfully compute the answer to any precision:

In[·]:= NSum[Sin[1/n]*Sin[n], {n, 1, Infinity},
        WorkingPrecision -> 30, Method -> "AlternatingSigns"]
Out[·]:= 0.986295119645840946771854264431

One can indicate the range of the terms in SumConvergence:

In[·]:= SumConvergence[1/n^2, {n, 0, Infinity}]
Out[·]:= False

In[·]:= SumConvergence[1/n^2, {n, 1, Infinity}]
Out[·]:= True

Mathematica® can decide on the convergence of the following series:

In[·]:= SumConvergence[(3^n + (-1)^n)^2/10^n, n]
Out[·]:= True

One can expand the terms as follows:

In[·]:= Expand[Refine[(3^n + (-1)^n)^2/10^n,
        Assumptions -> {Mod[n, 2] == 0}]]
Out[·]:= (3/5)^n*2^(1 - n) + 10^(-n) + 3^(2*n)/10^n

Note that NSum sometimes gives an incorrect answer:

In[·]:= NSum[((-1)^n/(3*n))*(2 + (-1)^n), {n, 1, Infinity}]
Out[·]:= 0.663773 - 0.0798565 I

Clearly the series above is divergent as the sum of a convergent and a divergent series. However Mathematica® fails to notice this. NSum returns an obviously false answer without any warning. NSum must verify convergence, since:

In[·]:= Options[NSum][[7]]
Out[·]:= VerifyConvergence → True

The problem is that SumConvergence cannot determine whether the sum of two series is convergent even when it can determine this for each term individually:

In[·]:= Refine[Expand[((-1)^n/(3*n))*(2 + (-1)^n)],
      Assumptions -> Element[n, PositiveIntegers]]
Out[·]:= 1/(3*n) + (2*(-1)^n)/(3*n)

In[·]:= Map[SumConvergence[#1, n] & , %, {1}]
Out[·]:= False + True

One can avoid this problem by using the option WorkingPrecision:

In[·]:= NSum[((-1)^n/(3*n))*(2 + (-1)^n), {n, 1,
      Infinity}, WorkingPrecision -> 100];

or the option NSumTerms with a large number of terms:

In[·]:= NSum[((-1)^n/(3*n))*(2 + (-1)^n), {n, 1,
      Infinity}, NSumTerms -> 10000];

In those two cases NSum issues warnings indicating that it has come across a problem with convergence (we omitted those warnings along with the outputs above).

3.3 Absolute convergence

We first introduce the notion of *absolute convergence*. We say that a series $\sum_{n=1}^{\infty} a_n$ is *absolutely convergent* if the series $\sum_{n=1}^{\infty} |a_n|$ is convergent. Note that when a is complex, then $|a|$ is the modulus of a, which is a positive real number.

The following properties hold.

1. If $\sum_{n=1}^{\infty} a_n$ is absolutely convergent, then $\sum_{n=1}^{\infty} a_n$ is convergent.
2. A convergent series with terms of constant sign (either positive or negative) is absolutely convergent. If $a_n > 0$ ($a_n < 0$) for all n, then $\sum_{n=1}^{\infty} a_n$ is convergent if and only if its sequence of partial sums is bounded above (below). When a series with positive (negative) terms is convergent we write $\sum_{n=1}^{\infty} a_n < \infty$ ($\sum_{n=1}^{\infty} a_n > -\infty$). We use this notation only for series whose terms have a constant sign.
3. If $\sum_{n=1}^{\infty} a_n$ is absolutely convergent and $\pi : \mathbb{N} \rightarrow \mathbb{N}$ is any bijection, then $\sum_{n=1}^{\infty} a_{\pi(n)} = \sum_{n=1}^{\infty} a_n$.

As we will see in Section 3.7, property 3 is not true for series which are not absolutely convergent.

The following tests can be used to show that the series is absolutely convergent. They can also sometimes be used to show that a series which is known to be not absolutely convergent is actually divergent.

D'Alembert's ratio test

Assume that $a_n \neq 0$. If $\lim_{n\to\infty} |a_{n+1}/a_n| = g \in \mathbb{R} \cup \{\infty, -\infty\}$, then the following statements hold.

- If $g < 1$, then the series $\sum_{n=1}^{\infty} a_n$ is absolutely convergent.
- If $g > 1$, then the series $\sum_{n=1}^{\infty} a_n$ is divergent.

Cauchy's root test

If $\lim_{n\to\infty} \sup |a_n|^{1/n} = g \in \mathbb{R} \cup \{\infty, -\infty\}$, then the following statements hold.

- If $g < 1$, then the series $\sum_{n=1}^{\infty} a_n$ is absolutely convergent.
- If $g > 1$, then the series $\sum_{n=1}^{\infty} a_n$ is divergent.

The root test is actually stronger than the ratio test: one can show that in all cases when the ratio test works, so does the root test, but one can construct examples in which d'Alembert's ratio test does not work but the Cauchy root test does (for instance, $\sum_{n=1}^{\infty} 2^{(-1)^n} 2^{-n}$). Note that the root test (and hence also the ratio test) fails when $g = 1$.

D'Alembert's ratio test and Cauchy's root test are implemented in the option Method in Mathematica®'s function SumConvergence, which we will discuss later on in this chapter.

3.4 Convergence of series with terms of constant signs

The theory of convergent series is traditionally divided into two parts. The first part is concerned with series whose terms have the same signs. The second part is concerned with series that have infinitely many of both positive and negative terms. Note that, just like in the case of a sequence, convergence of a series is unaffected by ignoring a finite number of terms. However, unlike the case of the limit of a sequence, removing a finite number of terms from a series may change the value of its sum (if that value is finite).

In this section we state without proof some basic facts about convergence of series with constant signs and describe several tests of convergence. The proofs can be found in [12] and [14].

The following two tests are extremely useful for proving convergence of series of positive terms by hand but are generally unsuitable for present day computer algebra programs. They reason is that they involve comparing a given series (which we are trying to test) with another one, about which we must already know whether it is convergent or not. Choosing a suitable series involves making an "educated guess", something that humans are still better at than computers.

The comparison test

Let a_n, $b_n > 0$ and suppose that $a_n \leq b_n$ for all $n \in \mathbb{N}$. Then the following statements hold.

- If $\sum_{n=1}^{\infty} b_n$ is convergent, then $\sum_{n=1}^{\infty} a_n$ is convergent.
- If $\sum_{n=1}^{\infty} a_n$ is divergent, then $\sum_{n=1}^{\infty} b_n$ is divergent.

The limit comparison test

Let a_n, $b_n > 0$ and suppose that $\lim_{n\to\infty} a_n/b_n = c$. Then the following statements hold.

- If $c > 0$, then $\sum_{n=1}^{\infty} b_n$ is convergent if and only if $\sum_{n=1}^{\infty} a_n$ is convergent (in this case the two sequences $\{a_n\}$ and $\{b_n\}$ are said to be similar).
- If $c = 0$, then the convergence of $\sum_{n=1}^{\infty} b_n$ implies the convergence of $\sum_{n=1}^{\infty} a_n$.
- If $c = \infty$, then the divergence of $\sum_{n=1}^{\infty} b_n$ implies the divergence of $\sum_{n=1}^{\infty} a_n$.

The next two tests are implemented in the option Method in Mathematica®'s function SumConvergence.

Raabe's test

Suppose $a_n > 0$ and $\lim_{n\to\infty} n(a_n/a_{n+1} - 1) = g \in \mathbb{R} \cup \{\infty, -\infty\}$. Then the following statements hold.

- If $g > 1$, then the series $\sum_{n=1}^{\infty} a_n$ is convergent.
- If $g < 1$, then the series $\sum_{n=1}^{\infty} a_n$ is divergent.

The last test uses the concept of an improper integral of a continuous function, which will be considered later. However, we shall state the test now (see also Chapter 7) because it is one of the most effective tests at Mathematica®'s disposal (due to the fact that Mathematica® is much better than an average human mathematician at integration).

Integral test

Let f be a continuous, positive, decreasing function of x for $x \geq 1$ and $\lim_{x\to\infty} f(x) = 0$. Then $\sum_{n=1}^{\infty} f(n)$ is convergent if and only if $\int_{1}^{\infty} f(x)dx$ is convergent.

The tests above are very useful for proving convergence of series of positive terms by hand, but are unsuitable for automatic use by current computer programs. The functionality of these tests was recently implemented in Mathematica® by introducing functions like AsymptoticLess and AsymptoticLessEqual. Note that there are other functions with "Asymptotic" in the name. For instance, one can verify that x is asymptotically less than x^2 ($x < x^2$) as $x \to \infty$:

In[·]:= AsymptoticLess[x, x^2, x -> Infinity]
Out[·]:= True

which means that $x \in o(x^2)$:

In[·]:= Limit[x/x^2, x -> Infinity]
Out[·]:= 0

These functions can be used to test series for convergence. For example, consider the series

$$\sum_{n=1}^{\infty} \frac{e^n}{e^{n+1/n} \ln^2 n}.$$

SumConvergence does not give an answer:

In[·]:= SumConvergence[Exp[n]/(Exp[n^(1 + 1/n)]*Log[n]^2), n]
Out[·]:= SumConvergence[E^(n - n^(1 + 1/n))/Log[n]^2, n]

We look at two terms of the series expansion at infinity:

In[·]:= Normal[Series[Exp[n]/(Exp[n^(1 + 1/n)]*Log[n]^2),
      {n, Infinity, 2}]]
Out[·]:= -(1/(2*n^2)) + 1/(n*Log[n]^2)

This suggests that we should compare the terms of our series with $1/(n \ln^2 n)$ using AsymptoticLessEqual:

In[·]:= AsymptoticLessEqual[Exp[n]/(Exp[n^(1 + 1/n)]*
      Log[n]^2), 1/(n*Log[n]^2), n -> Infinity]
Out[·]:= True

In[·]:= AsymptoticLess[Exp[n]/(Exp[n^(1 + 1/n)]*Log[n]^2),
      1/(n*Log[n]^2), n -> Infinity]
Out[·]:= False

Since

In[·]:= SumConvergence[1/(n*Log[n]^2), n]
Out[·]:= True

by the comparison test we can conclude that our series is convergent. Unfortunately, this functionality is still far from perfect and one can easily find examples which are easy to solve by hand but for which AsymptoticLessEqual fails to give an answer (sometimes after waiting for a long time):

In[·]:= SumConvergence[((2 + (-1)^n)/6)^n, n]
Out[·]:= SumConvergence[(2 + (-1)^n)^n/6^n, n]

In[·]:= AsymptoticLessEqual[((2 + (-1)^n)/6)^n,
      (1/2)^n, n -> Infinity]
Out[·]:= AsymptoticLessEqual[(2 + (-1)^n)^n/6^n,
      2^(-n), n -> Infinity]

In[·]:= AsymptoticEqual[((2 + (-1)^n)/6)^n,
      (1/2)^n, n -> Infinity]
Out[·]:= AsymptoticEqual[(2 + (-1)^n)^n/6^n,
      2^(-n), n -> Infinity]

Another example is the following:

In[·]:= SumConvergence[(n^2 - 100*n)/(n^4 + 1000*n^3), n]
Out[·]:= True

The terms of this series are positive for $n > 100$:

In[·]:= Reduce[(n^2 - 100*n)/(n^4 + 1000*n^3) > 0, n]
Out[·]:= n < -1000 || n > 100

so we can apply the asymptotic test and compare them with $1/n^2$:

In[·]:= AsymptoticEqual[(n^2 - 100*n)/(n^4 + 1000*n^3),
        1/n^2, n -> Infinity]
Out[·]:= True

Below are a few more examples.

In[·]:= SumConvergence[2^(1/n) - 1, n]
Out[·]:= False

In[·]:= AsymptoticEqual[2^(1/n) - 1, 1/n,
        n -> Infinity]
Out[·]:= True

In[·]:= AsymptoticGreaterEqual[(1 - Log[n]/n)^n,
        1/n, n -> Infinity]
Out[·]:= True

In[·]:= AsymptoticLessEqual[1/n^Log[Log[n]],
        1/n^2, n -> Infinity]
Out[·]:= True

Using Reduce we can compute exactly when the terms of the first series become smaller than the corresponding terms of the second series:

In[·]:= Reduce[1/n^Log[Log[n]] <= 1/n^2, n,
        PositiveIntegers]
Out[·]:= Element[n, Integers] && n >= 1619

It is also possible to use assumptions:

In[·]:= AsymptoticLess[n^b*Log[n]^c, n, n ->
        Infinity, Assumptions -> b < 1]
Out[·]:= True

3.4.1 Example

Let us study for which values of the parameter a the following series $\sum_{n=1}^{\infty}(5^{1/n} - 1)^a$ is convergent.

In[·]:= SumConvergence[(5^(1/n) - 1)^a, n, Assumptions ->
        {Element[a, Reals]}]
Out[·]:= a > 1

Let us try to use the comparison test. We will see later that it is enough to deal with the case $a = 1$ so let us consider it first. The trick is to choose the right series to compare. It is often useful to look at some graphs. We would like to find two positive integers p, q and constants c_1 and c_2 such that the graph of $5^{1/n} - 1$ lies between that of c_1/n^p and c_2/n^q for sufficiently large n. We could use Manipulate to find candidates for such p and q but we will illustrate this only with the usual "static" graph:

In[·]:= `DiscretePlot[{5^(1/n) - 1, 1/n, 3/n}, {n, 1, 20},`
        `PlotMarkers -> {Automatic, 9}]`

Out[ ]=

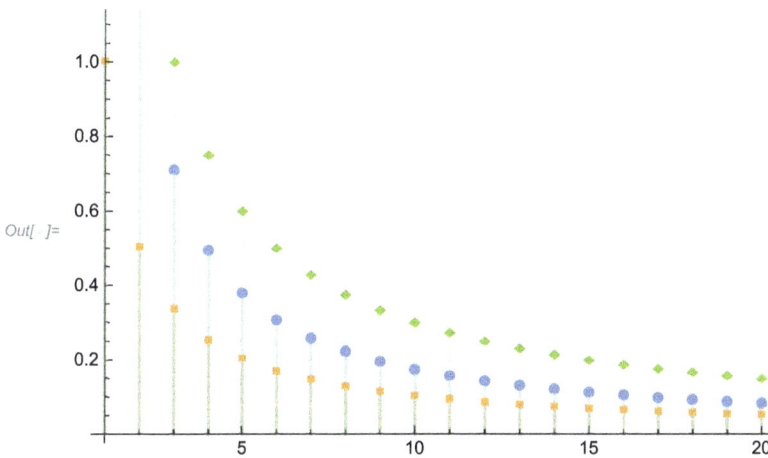

Figure 3.1

So the graph suggests that $1/n < 5^{1/n} - 1 < 3/n$ for all $n > 1$, which we can try to verify by the function Reduce:

In[·]:= `Reduce[1/n < 5^(1/n) - 1 < 3/n && n > 1, n, Integers]`
Out[·]:= $n \in \mathbb{Z} \,\&\&\, n \geq 2$

Hence, $1/n^a < (5^{1/n} - 1)^a < 3^a/n^a$ and the series $\sum_{n=1}^{\infty}(5^{1/n} - 1)^a$ is convergent for $a > 1$ and divergent for $a \leq 1$.

The limit comparison test works even quicker since we do not need to bother trying to find suitable p and q. However, this requires the ability to compute limits:

In[·]:= `DiscreteLimit[(5^(1/n) - 1)^a/(1/n^a), n -> Infinity,`
        `Assumptions -> {a > 0}]`
Out[·]:= $\text{Log}[5]^a$

We shall see how this limit can be computed by hand later, when we consider the Taylor series. However, somewhat surprisingly the inequalities we use to apply the comparison test can be proved rather easily (which is again not how Mathematica® proves them).

Consider the inequality $5^{1/n}-1 < 3/n$. Transforming we easily see that it is equivalent to $5 < (1+3/n)^n$. Now let us write the right hand side as $((1+1/(n/3))^{n/3})^3$. We have already seen that this sequence is increasing (in fact its limit is e^3). For $n = 3$ its n-th term is 8, hence it and all the others are larger than 5 (actually even for $n = 2$ the inequality already holds but we only need to show that it holds for some n).

Next let us consider the inequality $1/n < 5^{1/n} - 1$. It is equivalent to $(1+1/n)^n < 5$. We know that the sequence on the left hand side is monotonically increasing and is always less than 3. Hence it is less then 5.

3.4.2 Example

Let us find the convergent series with positive terms a_n for which na_n does not tend to zero.

We define a_n as $1/n$ when n is a square of a natural number and $1/n^2$ otherwise. The terms of such a series are as follows:

In[·]:= ls = Table[If[Element[Sqrt[n], Integers],
            1/n, 1/n^2], {n, 1, 6}]
Out[·]:= {1, 1/4, 1/9, 1/4, 1/25, 1/36}

This sequence of terms consists of reciprocals of all squares, hence, the corresponding series is convergent. However, the sequence na_n does not have a limit since at each position where n is a square, na_n is equal to 1, as can be seen below:

In[·]:= MapIndexed[#2[[1]]*#1 & , ls]
Out[·]:= {1, 1/2, 1/3, 1, 1/5, 1/6}

3.4.3 Example

Let a_n satisfy the recurrence $a_{n+1} = 1 + 1/a_n$, $a_1 = 1$. Find $g = \lim_{n\to\infty} a_n$ and determine if the series $\sum_{n=1}^{\infty} |a_n - g|$ converges.

Note that the recurrence was considered in Section 2.2. Recall that the solution was given by

In[·]:= FullSimplify[a[n] /. RSolve[{a[n + 1] ==
            1 + 1/a[n], a[1] == 1}, a[n], n][[1]]]
Out[·]:= 1/2 + LucasL[n]/(-Fibonacci[1 + n] + LucasL[1 + n])

and its limit by

In[·]:= g = Limit[%, n -> Infinity]
Out[·]:= (1/2)*(1 + Sqrt[5])

We want to investigate the convergence of the series $\sum_{n=1}^{\infty} |a_n - g|$. We will use the d'Alembert (ratio) criterion. That is, we want to compute

$$\lim_{n\to\infty} \frac{|a_{n+1} - g|}{|a_n - g|}.$$

Substituting for a_{n+1} from the recursive formula, this is the same as:

```
In[·]:= Limit[Abs[1 + 1/an - g]/Abs[an - g], an -> g]
Out[·]:= (1/2)*(3 - Sqrt[5])
```

```
In[·]:= N[(1/2)*(3 - Sqrt[5])]
Out[·]:= 0.381966
```

This is less than 1, so by the ratio test the series converges.

3.5 Convergence of series with terms of non-constant signs

In this section we will discuss some methods of proving convergence of series whose terms do not have a constant sign.

3.5.1 Grouping of terms

A simple method, which is useful when dealing with convergence of series with non-constant signs is a method of grouping of terms. Namely, consider the relationship between the following two series:

$$\sum_{i=0}^{\infty} a_i = a_0 + a_1 + a_2 + a_3 + \cdots \qquad (3.4)$$

and

$$\sum_{i=0}^{\infty} (a_{2i+1} + a_{2i}) = (a_0 + a_1) + (a_2 + a_3) + \cdots \qquad (3.5)$$

First note that if (3.4) is convergent, so is (3.5). This can be seen by looking at the sequences of partial sums and observing that the second sequence is a subsequence of the first. Since any subsequence of a convergent sequence is convergent, the result follows. Now suppose that we know that the second series is convergent. Then it does not in general follow that the first also is. For example, consider the series $(1-1) + (1-1) + \cdots$. This is just the series $0 + 0 + \cdots$, so its partial sums are all 0 and the limit of the (constant) sequence of zeros is 0. But the series $1 - 1 + 1 - 1 + \cdots = \sum_{n=0}^{\infty} (-1)^n$ is divergent, since its sequence of partial sums is 1, 0, 1, ...

However, suppose now we know that (3.5) is convergent and also that the sequence $\{a_n\}$ is a null sequence (that is, the necessary condition for convergence of (3.4) is satisfied). In this case from the convergence of (3.5) follows the convergence of (3.4). Indeed, consider the sequence of partial sums of (3.4) $\{s_n\}$, where $s_n = \sum_{i=0}^{n} a_i$. We have already seen that a sequence is convergent if and only if its two subsequences of even indexed terms and odd indexed terms are convergent to the same limit (Lemma 4 in Chapter 2). The sequence $\{s_{2n}\}$ is just the sequence of terms of (3.5), hence it is convergent. The sequence $\{s_{2n+1}\}$ also converges to the same limit because $s_{2n+1} = s_{2n} + a_{2n+1}$ and $a_{2n+1} \to 0$. (We see that we only needed the assumption that $a_{2n+1} \to 0$ rather than $a_n \to 0$.)

Informally speaking we have shown that if the necessary condition for convergence of a series, i. e., the condition $a_n \to 0$, is satisfied, we can "group" and "ungroup" the series into pairs without affecting convergence. The same argument shows that this is also true about grouping the terms (or "inserting brackets") into groups of terms of arbitrary fixed length (e. g., 3): $(a_0 + a_1 + a_2) + (a_3 + a_4 + a_5) + (a_6 + a_7 + a_8) + \cdots$. In fact, it can even be proved that we can group a series into segments of various lengths, as long as their lengths are bounded, e. g., all less than 30. However, we shall not need these results here.

In Mathematica® one can "insert brackets". Such groupings can be displayed by using some of the already described methods of preventing evaluation. For example, consider the alternating harmonic series:

In[·]:= `Inactivate[Sum[(-1)^(n + 1)/n, {n, 1,`
        `Infinity}], Sum]`
Out[·]:= `Inactive[Sum][(-1)^(1 + n)/n, {n, 1, Infinity}]`

We can group the terms of this series into pairs and display several of them:

In[·]:= `Sum[plus[1/(2*n - 1), -(2*n)^(-1)],`
        `{n, 1, 4}] /. plus[x_, y_] :> HoldForm[x + y]`
Out[·]:= $\left(1 - \frac{1}{2}\right) + \left(\frac{1}{3} + \frac{1}{4}\right) + \left(\frac{1}{5} + \frac{1}{6}\right) + \left(\frac{1}{7} + \frac{1}{8}\right)$

Alternatively we can do it using `Inactivate`:

In[·]:= `Reverse /@ Inactivate[Sum[1/Activate[2*n - 1]`
        `- 1/(2*n), {n, 1, 4}], Plus]`
Out[·]:= $\left(-\frac{1}{2} + 1\right) + \left(-\frac{1}{4} + \frac{1}{3}\right) + \left(-\frac{1}{6} + \frac{1}{5}\right) + \left(-\frac{1}{8} + \frac{1}{7}\right)$

Note that the terms of the grouped series (after adding the ones inside brackets) are positive so one can use the comparison or the asymptotic criterion.

3.5.2 Example

Let us show that the series $\sum_{n=1}^{\infty} (-1)^{n+1}/\sqrt{n}$ is convergent. We simply group the terms in pairs and observe that

In[·]:= `Refine[Together[(-1)^(2*k - 1)/Sqrt[2*k - 1]`
        `+ (-1)^(2*k)/Sqrt[2*k]], Element[k, Integers]]`
Out[·]:= $\dfrac{-2\sqrt{k} + \sqrt{2}\sqrt{-1 + 2k}}{2\sqrt{k}\sqrt{-1 + 2k}}$

This expression is always negative. It can actually be reduced to a "simpler" form by "rationalizing" the numerator. Mathematica®'s functions Simplify or FullSimplify will not do this automatically, but we already know that we can try to change the default ComplexityFunction as follows:

In[·]:= g[k_]:=FullSimplify[(Sqrt[2]*Sqrt[2*k - 1] - 2*
        Sqrt[k])/(2*Sqrt[k]*Sqrt[2*k - 1]), ComplexityFunction
        -> Function[x, Count[Numerator[x], _Power, Infinity]]]

In[·]:= g[k]

$$Out[·]:= \frac{1}{\sqrt{k}(\sqrt{2} - 2\sqrt{2}k - 2\sqrt{k}\sqrt{-1 + 2k})}$$

In[·]:= Simplify[Sign[g[k]], Assumptions ->
        {Element[k, Integers], k > 0}]

Out[·]:= -1

Since

In[·]:= DiscreteLimit[g[k]/(1/k^(3/2)), k -> Infinity]

$$Out[·]:= -\frac{1}{4\sqrt{2}}$$

the series is convergent by the limit comparison test.

3.5.3 Example

Show that the series

$$\frac{1}{3} + \frac{1}{5} - \frac{1}{2} + \frac{1}{7} + \frac{1}{9} - \frac{1}{4} + \frac{1}{11} + \frac{1}{13} - \frac{1}{6} \cdots$$

is convergent and find its sum.

We shall attempt to write the series as $\sum_{n=1}^{\infty} a_n$, where a_n is some rational function of n. The series is

In[·]:= Flatten[Table[{1/(4*n - 1), 1/(4*n + 1),
        -(2*n)^(-1)}, {n, 1, 3}]]

Out[·]:= {1/3, 1/5, -1/2, 1/7, 1/9, -1/4, 1/11, 1/13, -1/6}

Since $a_n \to 0$ as $n \to \infty$, we can try to group three terms together. Then the general term of the new series will be

In[·]:= Together[Plus @@ {1/(4*n - 1), 1/(4*n + 1),
        -(2*n)^(-1)}]

Out[·]:= 1/(2*n*(-1 + 4*n)*(1 + 4*n))

The expression above is positive. The ratio test does not give a definite answer in this case. We can use the built-in function DiscreteRatio for this:

```
In[·]:= DiscreteRatio[f[n], n]
Out[·]:= f[1 + n]/f[n]
```

```
In[·]:= Limit[Simplify[DiscreteRatio[1/(2*n*(-1 + 4*n)*
        (1 + 4*n)), n]], n -> Infinity]
Out[·]:= 1
```

However, the expression above is asymptotically equal to $1/n^3$:

```
In[·]:= AsymptoticEqual[1/(2*n*(-1 + 4*n)*
        (1 + 4*n)), 1/n^3, n -> Infinity]
Out[·]:= True
```

Hence, the series is convergent. The sum is

```
In[·]:= PowerExpand[Sum[1/(2*n*(-1 + 4*n)*(1 + 4*n)),
        {n, 1, Infinity}]]
Out[·]:= (1/2)*(-2 + 3*Log[2])
```

3.5.4 Abel's summation formula

First, recall the definition of the inner or dot product. We can think of it as the product of two lists of the same length (or vectors of the same dimension):

```
In[·]:= l[a_, n_] := Table[a[i], {i, 1, n}]
In[·]:= l[a, 3]
Out[·]:= {a[1], a[2], a[3]}
```

```
In[·]:= l[a, 3] . l[b, 3]
Out[·]:= a[1] b[1] + a[2] b[2] + a[3] b[3]
```

```
In[·]:= Dot[l[a, 3], l[b, 3]]
Out[·]:= a[1] b[1] + a[2] b[2] + a[3] b[3]
```

In fact, Mathematica®'s function Dot is a special case of a more general function called Inner, but as we will not need it here, we will not say any more about it.

It is important not to omit the dot, because otherwise one gets a very different product:

```
In[·]:= l[a, 3] l[b, 3]
Out[·]:= {a[1] b[1], a[2] b[2], a[3] b[3]}
```

The dot is also necessary when multiplying matrices; without it we get a very different answer, which is not the usual product of matrices:

```
In[·]:= {{a, b}, {c, d}}.{{e, f}, {g, h}}
Out[·]:= {{a e + b g, a f + b h}, {c e + d g, c f + d h}}
```

In[·]:= {{a, b}, {c, d}} {{e, f}, {g, h}}
Out[·]:= {{a e, b f}, {c g, d h}}

We will now consider two lists, such that the length of the first one is one less than the length of the second one, say, 4 and 5:

In[·]:= list1 = l[a, 4]
Out[·]:= {a[1], a[2], a[3], a[4]}

In[·]:= list2 = l[b, 5]
Out[·]:= {b[1], b[2], b[3], b[4], b[5]}

Note that

In[·]:= Length[Differences[list2]]
Out[·]:= 4

In[·]:= Length[Accumulate[list1]]
Out[·]:= 4

This means that it makes sense to consider the dot product of Differences[list2] and Accumulate[list1]:

In[·]:= Collect[Expand[Differences[list2] .
          Accumulate[list1]], a[_]]
Out[·]:= a[1] (-b[1] + b[5]) + a[2] (-b[2] + b[5]) +
          a[3] (-b[3] + b[5]) + a[4] (-b[4] + b[5])

It suggests the following formula:

In[·]:= Simplify[Differences[list2] . Accumulate[list1]
          == -Most[list2] . list1 + Total[list1]*Last[list2]]
Out[·]:= True

Rearranging and rewriting in usual mathematical (rather then in Mathematica®'s) notation we get *Abel's summation formula*, which is easy to prove by induction:

$$\sum_{i=1}^{n} a_i b_i = \sum_{k=1}^{n} \left(\sum_{i=1}^{k} a_i \right) (b_k - b_{k+1}) + b_{n+1} \sum_{i=1}^{n} a_i. \tag{3.6}$$

3.5.5 Dirichlet's and Abel's tests

Suppose that we are given two sequences $\{a_n\}$ and $\{b_n\}$. What are the weakest conditions that we need to impose on both so that we can conclude that the "inner product" (or the "dot product") $\sum_{n=1}^{\infty} a_n b_n$ is convergent? For example, if $a_n > 0$ and $b_n > 0$ for all n, and $\sum_{n=1}^{\infty} a_n < \infty$, $\sum_{n=1}^{\infty} b_n < \infty$, then it is easy to see that $\sum_{n=1}^{\infty} a_n b_n < \infty$. Indeed, for a series of positive terms, convergence is equivalent to boundedness of its sequence of partial sums. The conclusion follows from the inequality $\sum_{i=1}^{m} a_i b_i \leq (\sum_{i=1}^{m} a_i)(\sum_{i=1}^{m} b_i)$, which follows simply by expanding the right hand side. But the result is not true in general. For example, let us take $a_n = (-1)^{n+1}/\sqrt{n}$, $b_n = (-1)^{n+1}/\sqrt{n}$. We have

In[·]:= SumConvergence[(-1)^(n + 1)*(1/Sqrt[n]), n]
Out[·]:= True

In[·]:= SumConvergence[(-1)^(n + 1)*(1/Sqrt[n])*(-1)^(n + 1)*
(1/Sqrt[n]), n]
Out[·]:= False

The second result is just the fact that the harmonic series is divergent.

We can think of Abel's summation formula (3.6) as an identity involving sequences. On the left hand side we have the sequence of partial sums of the sequence $\{a_i b_i\}$, on the right hand side there is the sum of two sequences. Thus, if we find a set of conditions that make both sequences on the right hand side converge, so will the sequence on the left.

Consider first the sequence $\sum_{k=1}^{n}(\sum_{i=1}^{k} a_i)(b_k - b_{k+1})$ (the dot product of the sequence of partial sums of one sequence and the difference sequence of the other). We will always assume that the sequence $\{b_n\}$ is monotone, so the differences all have the same sign. Since $\sum_{k=1}^{n}(b_k - b_{k+1}) = b_1 - b_{n+1}$, this is convergent if $\lim_{n\to\infty} b_n \in \mathbb{R}$. Since the series of differences has constant signs it is absolutely convergent. Now suppose that all the partial sums $\sum_{i=1}^{k} a_i$ are bounded by M. Then

$$\sum_{k=1}^{n}\left|\left(\sum_{i=1}^{k} a_i\right)(b_k - b_{k+1})\right| \leq \sum_{k=1}^{n} M|b_k - b_{k+1}| = M\sum_{k=1}^{n}|b_k - b_{k+1}|,$$

which is bounded. Hence $\sum_{k=1}^{n}(\sum_{i=1}^{k} a_i)(b_k - b_{k+1})$ is absolutely convergent. Now consider the other term $b_{n+1}\sum_{i=1}^{n} a_i$. From the fact that the sequence $\{b_{n+1}\}$ is convergent and $\sum_{i=1}^{n} a_i$ is bounded, it does not follow that $b_{n+1}\sum_{i=1}^{n} a_i$ is convergent. We need to make a stronger assumption. One possibility is to assume that $\lim_{n\to\infty} b_n = 0$ and $\sum_{i=1}^{n} a_i$ are bounded. In this case the term $b_{n+1}\sum_{i=1}^{n} a_i$ also converges (to 0) and we can conclude that $\sum_{i=1}^{n} a_i b_i$ converges. If we only assume that $\lim_{n\to\infty} b_n = g \in \mathbb{R}$, but g is not necessarily 0, we need a stronger assumption about $\{a_n\}$, namely, that the sequence $\sum_{i=1}^{n} a_i$ converges, that is, the series $\sum_{i=1}^{\infty} a_i$ is convergent. Thus we have proved Dirichlet's and Abel's criteria (tests) for the convergence of inner product of sequences.

Theorem 7. *Suppose we have two sequences $\{a_n\}$ and $\{b_n\}$, where $\{b_n\}$ is monotone convergent to g. Suppose also that one of the following conditions holds.*

Dirichlet's test
The partial sums $\sum_{i=1}^{n} a_i$ are bounded and $g = 0$.

Abel's test
The series $\sum_{i=1}^{\infty} a_i$ is convergent.
Then $\sum_{i=1}^{\infty} a_i b_i$ is convergent.

Its easy to show that one can deduce the Abel test from the Dirichlet test. One well-known consequence of Abel's test is the Leibniz test for convergence.

Leibniz's test

Let $\{c_n\}$ be a monotonic null sequence (i. e., $\lim_{n\to\infty} c_n = 0$). Then $\sum_{n=0}^{\infty}(-1)^n c_n$ converges.

Indeed, we simply take in the statement of the Dirichlet test $a_n = (-1)^n$ and $b_n = c_n$. Clearly, $\sum_{i=1}^{n} a_i$ are bounded and $\{b_n\}$ is a monotonic null sequence.

The Dirichlet test is, of course, more general than the Leibniz test. For example, it applies to examples such as

In[·]:= SumConvergence[(-1)^(n*((n + 1)/2))/n, n]
Out[·]:= True

In[·]:= SumConvergence[(-1)^Quotient[n, 5]/n, n]
Out[·]:= False

However,

In[·]:= SumConvergence[(-1)^n^2015/n, n]

$$Out[\cdot]:= \text{SumConvergence} \left[\frac{(-1)^{n^{2015}}}{n}, n \right]$$

Here we need to observe something that Mathematica® fails to notice (this could be changed in future versions), namely, that

In[·]:= FullSimplify[Mod[n^2015, 2], Element[n, Integers]]
Out[·]:= Mod[n, 2]

That means that Mathematica® should simplify $(-1)^{n^{2015}}/n$ to $(-1)^n/n$. That it does not do so in version 14.2 we consider an omission. In each case the series can be thought of as a "dot product" of two sequences satisfying the conditions of the Dirichlet test. Sometimes such a decomposition and particularly the fact that one of the sequences has bounded partial sums can be difficult to notice.

3.5.6 Example

Consider $\sum_{n=1}^{\infty} \sin(nx)/n$ and suppose that x is not an integer multiple of π (otherwise all the terms are zero). Testing for convergence we get:

In[·]:= SumConvergence[Sin[n*x]/n, n, Assumptions ->
         Element[x, Reals]]
Out[·]:= True

It is not obvious that one can use the Dirichlet test. The series can be regarded as the dot product of two sequences $\{1/n\}$ and $\{\sin(nx)\}$, but does the latter have bounded partial sums? We can check this with Mathematica®:

In[·]:= Sum[Sin[n*x], {n, 1, m}]

$$Out[\cdot]:= \text{Csc}\left[\frac{x}{2}\right] \text{Sin}\left[\frac{m\,x}{2}\right] \text{Sin}\left[\frac{1}{2}(1 + m)x\right]$$

Since

$$\left| \csc\left(\frac{x}{2}\right) \sin\left(\frac{mx}{2}\right) \sin\left(\frac{1}{2}(m+1)x\right) \right| \le \left| \csc\left(\frac{x}{2}\right) \right| = \left| \frac{1}{\sin(x/2)} \right|$$

is independent of m, the partial sums are bounded for fixed x.

The trigonometric identity

$$\sum_{n=1}^{m} \sin(nx) = \csc\left(\frac{x}{2}\right) \sin\left(\frac{mx}{2}\right) \sin\left(\frac{1}{2}(m+1)x\right)$$

can be proved by purely trigonometric means (we leave this as an exercise for those readers interested in elementary trigonometry) but the easiest way is by using Euler's formula in complex analysis relating the exponential and trigonometric functions. This can be obtained in Mathematica® in several ways, e. g.,

In[·]:= ComplexExpand[Exp[I*x]]
Out[·]:= Cos[x] + I Sin[x]

In[·]:= ExpToTrig[Exp[I*x]]
Out[·]:= Cos[x] + I Sin[x]

We now see that

$$\sum_{n=1}^{m} \sin(nx) = \text{Im}\left(\sum_{n=1}^{m} (\cos(nx) + i\,\sin(nx)) \right) = \text{Im}\left(\sum_{n=1}^{m} e^{inx} \right).$$

Clearly,

$$\sum_{n=1}^{m} e^{inx} = \frac{e^{ix}(e^{imx} - 1)}{e^{ix} - 1}$$

as the sum of m terms of a geometric progression. Hence,

In[·]:= Simplify[ComplexExpand[Im[(E^(I*x)*(-1 + E^(I*m*x)))/
        (-1 + E^(I*x))]]]
Out[·]:= Csc$\left[\frac{x}{2}\right]$ Sin$\left[\frac{m x}{2}\right]$ Sin$\left[\frac{1}{2}(1 + m)x\right]$

Note that Mathematica® can actually find the sum of this series:

In[·]:= Sum[Sin[n*x]/n, {n, 1, Infinity}]
Out[·]:= $\frac{1}{2}$ I (Log$[1 - E^{Ix}]$ − Log$[E^{-Ix}(-1 + E^{Ix})]$)

Mathematica®'s answer unfortunately involves the complex i, which is inconvenient for some purposes. Even if we use FullSimplify with the assumption that x is real, we get the same answer:

In[·]:= `Assuming[Element[x, Reals], FullSimplify[Sum[Sin[n*x]/n,`
        `{n, 1, Infinity}]]]`

Out[·]:= $\dfrac{1}{2} I \left(\text{Log}\left[1 - E^{Ix}\right] - \text{Log}\left[E^{-Ix}\left(-1 + E^{Ix}\right)\right]\right)$

However, since we know that the sum of real numbers is surely real we can try to get an answer not involving complex *i* in a different way:

In[·]:= `FullSimplify[ComplexExpand[Re[Sum[Sin[n*x]/n,`
        `{n, 1, Infinity}]], TargetFunctions -> {Im, Re}]]`

Out[·]:= $\dfrac{1}{2}$ `(-ArcTan[1 - Cos[x], -Sin[x]] +`
        `ArcTan[1 - Cos[x], Sin[x]])`

The answer involves a two-argument version of the inverse trigonometric function $\tan^{-1}$ but involves no explicit *i*. Moreover, we can similarly show that

In[·]:= `FullSimplify[ComplexExpand[Im[Sum[Sin[n*x]/n,`
        `{n, 1, Infinity}]], TargetFunctions -> {Im, Re}]]`

Out[·]:= `0`

One should be aware that Dirichlet's and Abel's tests are not really suitable for present day computer programs since they require "guessing" how to decompose an infinite series into a dot product of two sequences with the required properties. For example, the current version of Mathematica® cannot determine the convergence of the series $\sum_{n=1}^{\infty} \sin(n)\sin(1/n)$ even though it is very similar to the above one:

In[·]:= `SumConvergence[Sin[1/n]*Sin[n], n]`

Out[·]:= `SumConvergence` $\left[\text{Sin}\left[\dfrac{1}{n}\right]\text{Sin}[n], n\right]$

We know that {sin(n)} has bounded partial sums and {sin(1/n)} monotonically converges to 0, so to human eyes the problem of convergence is not harder than before.

3.6 The function SumConvergence

This section will be devoted to the deeper study of the function SumConvergence.

In[·]:= `Options[SumConvergence]`

Out[·]:= `{Assumptions :> $Assumptions, Direction -> 1,`
        `Method -> Automatic}`

The option Direction gives us the direction of summation, where the indices can go to ∞ or to $-\infty$:

In[·]:= `SumConvergence[1/z^n, n, Direction -> 1]`

Out[·]:= `Abs[z] > 1`

In[·]:= `SumConvergence[1/z^n, n, Direction -> -1]`

Out[·]:= `Abs[z] < 1`

For us the most interesting option of SumConvergence is Method. It takes four possible values: "IntegralTest", "RaabeTest", "RatioTest", "RootTest". These are not the only tests that Mathematica® uses but they are the only ones that the user can choose. When we tell Mathematica® to use one of these tests, it will return the answer True or False depending on the result of the test. Let us first look at some examples:

```
In[·]:= tests = {"IntegralTest", "RaabeTest", "RatioTest",
        "RootTest"};
In[·]:= (SumConvergence[1/n^n, n, Method -> #1] & ) /@ tests
Out[·]:= {SumConvergence[n⁻ⁿ, n, Method -> "IntegralTest"],
        True, True, True}
```

We see that in this case all the tests except for the integral test succeeded.

Note also that the function SumConvergence can also be used with series involving a parameter and that one can use the assumptions mechanism. For example,

```
In[·]:= SumConvergence[n^a, n, Assumptions -> Element[a,
        Reals]]
Out[·]:= 1+a < 0
```

```
In[·]:= SumConvergence[(n^(1/n) - 1)^a, n,
        Assumptions -> Element[a, Reals]]
Out[·]:= a > 1
```

Without using Assumptions Mathematica® cannot give an answer:

```
In[·]:= SumConvergence[(n^(1/n) - 1)^a, n]
```

$$Out[·]:= \text{SumConvergence}\left[\left(-1 + n^{\frac{1}{n}}\right)^a, n\right]$$

We can check that none of these tests in our list will work for the alternating harmonic series:

```
In[·]:= (SumConvergence[(-1)^(n + 1)*(1/n),
        n, Method -> #1] & ) /@ tests
```

$$Out[·]:= \{\text{SumConvergence}\left[\frac{(-1)^{1+n}}{n}, n, \text{Method} \to \text{IntegralTest}\right],$$
$$\text{SumConvergence}\left[\frac{(-1)^{1+n}}{n}, n, \text{Method} \to \text{RaabeTest}\right],$$
$$\text{SumConvergence}\left[\frac{(-1)^{1+n}}{n}, n, \text{Method} \to \text{RatiolTest}\right],$$
$$\text{SumConvergence}\left[\frac{(-1)^{1+n}}{n}, n, \text{Method} \to \text{RootTest}\right]\}$$

Nevertheless Mathematica® can show convergence:

```
In[·]:= SumConvergence[(-1)^(n + 1)*(1/n), n]
Out[·]:= True
```

3.6.1 Example

Let us consider convergence of the series of the form $\sum_{n=1}^{\infty} n^{-p}$:

In[·]:= SumConvergence[n^(-p), n]
Out[·]:= Re[p] > 1

Mathematica®'s answer is actually valid for complex p. If we wanted to consider only real values we could have used the following expression:

In[·]:= SumConvergence[n^(-p), n, Assumptions ->
      Element[p, Reals]]
Out[·]:= p > 1

Let us check which of the tests was successful:

In[·]:= (SumConvergence[n^(-p), n, Method -> #1,
      Assumptions -> Element[p, Reals]] &) /@ tests
Out[·]:= {p > 1, p > 1, SumConvergence[n^$^{-p}$, n, Method ->
      RatioTest, Assumptions -> p ∈ ℝ],
      SumConvergence[n^{-p}, n, Method ->
      RootTest, Assumptions -> p ∈ ℝ]}

We see that both the integral test and Raabe's test work. We can verify this with Mathematica®:

In[·]:= Integrate[x^(-p), {x, 1, Infinity},
      Assumptions -> Element[p, Reals]]
Out[·]:= ConditionalExpression$\left[\dfrac{1}{-1+p}, p > 1\right]$

In[·]:= Limit[n*((n + 1)^p/n^p - 1), n -> Infinity]
Out[·]:= p

Both Raabe's test and the integral test show that the series is convergent when $p > 1$. Raabe's test also shows that the series is divergent for $p < 1$.

The series $\sum_{n=1}^{\infty} 1/n$ is known as the harmonic series. It is divergent and provides an example of a series which satisfies the condition that its n-th term tends to 0 yet is not convergent. The series $\sum_{n=1}^{\infty} 1/n^p$ for various p provide a very useful family which very often turn out to be suitable for using in the comparison or the limit comparison tests.

Let us now consider another problem that Mathematica® is unable to solve: determine if the series

$$\sum_{n=1}^{\infty} \left(\frac{1}{n^{9/8}} + \frac{(-1)^n}{n} \right) \tag{3.7}$$

is convergent.

In[·]:= SumConvergence[1/n^(9/8) + (-1)^n/n, n]

Out[·]:= SumConvergence $\left[\dfrac{1}{n^{9/8}} + \dfrac{(-1)^n}{n}, n \right]$

Of course, it is very easy to see that this series is convergent. We simply use the well-known and easy to prove fact (which follows from the earlier stated property of limits of sequences) that the sum of the sum of two series is the sum of their sums, provided everything is defined.

In[·]:= SumConvergence[1/n^(9/8), n]
Out[·]:= True

In[·]:= SumConvergence[(-1)^n/n, n]
Out[·]:= True

Hence, the series (3.7) is convergent.

3.7 Riemann's theorem on conditionally convergent series

We now know that the alternate harmonic series $\sum_{n=1}^{\infty} (-1)^{n+1}/n$ is convergent although it is not absolutely convergent. Such a series is called conditionally convergent. Conditionally convergent series are tricky as their sums do no behave like ordinary sums. For example, these sums (that is, the limits of the sequences of partial sums) can change when the terms are rearranged. In fact, *Riemann's theorem* on conditionally convergent series [14, Section 5.3.3] asserts that for any $a \in \mathbb{R} \cup \{\infty, -\infty\}$ one can find a permutation of terms such that the sum of the series will be a (that includes the cases $a = \infty$ and $a = -\infty$). The proof is an algorithm which is easy to implement in Mathematica®. This is done in [17], where some interesting related questions are considered. Here we will give a different implementation which is more suitable for a graphic representation of the theorem.

Consider for instance the alternate harmonic series $\sum_{n=1}^{\infty} (-1)^{n+1}/n$ and assume that $a \in \mathbb{R}$. The algorithm is as follows. Let A be the set of all positive terms of this series and let B be the set of all negative terms of this series, all arranged in the order of the appearance in the original series. We start by adding elements of A in turn until we reach a number a_1 which is greater than a. Because the sum of the series consisting of elements of A is infinity, this must always occur after a finite number of additions. Then we start adding elements of B until our sum becomes $a_2 < a$. Then we again add a finite number of the remaining elements of A until the sum becomes $a_3 > a$ and we continue in this way. One can easily see that $|a - a_n| \to 0$, hence the sequence of partial sums a_n of the rearranged alternate harmonic series tends to a. The reader should supply an algorithm for $a = \infty$ $(-\infty)$.

The algorithm written above constructs partial sums that are arbitrary close to the given number a but it never stops. Of course, when we want to implement the algorithm on a computer, we have to decide when to stop. There are several natural places to do that, for example, when we obtain a sufficiently close approximation or when we have used a given number of terms of the series. Our implementation uses a slightly different approach. We provide a list of positive terms and negative terms and the program stops when it can no longer continue (when there are no more needed elements in one of the lists). The program returns a fragment of the rearranged sequence of terms of the alternate harmonic series (so that the partial sums of the series are obtained by applying the function Accumulate to the list) and approximate sum. Note that our algorithm works for an arbitrary conditionally convergent series.

```
In[·]:= Rearrangement[a_, pos_, neg_] :=
        Module[{p = pos, n = neg, s = {}, sum = 0},
        Catch[While[p != {} || n != {},
        While[sum <= a, If[p == {}, Throw[{s, N[sum]}],
        AppendTo[s, First[p]]; sum = sum + First[p];
        p = Rest[p]]]; While[sum >= a, If[n == {},
        Throw[{s, N[sum]}], AppendTo[s, First[n]];
        sum = sum + First[n]; n = Rest[n]]]]; {s, N[sum]}]]
```

For the alternate harmonic series we have:

```
In[·]:= listpos[n_] := Table[(-1)^(k + 1)/k, {k, 1, n, 2}];
In[·]:= listneg[n_] := Table[(-1)^(k + 1)/k, {k, 2, n, 2}];
In[·]:= Rearrangement[2, listpos[30], listneg[20]]
```

$$Out[·]:= \left\{\left\{1, \frac{1}{3}, \frac{1}{5}, \frac{1}{7}, \frac{1}{9}, \frac{1}{11}, \frac{1}{13}, \frac{1}{15}, -\frac{1}{2}, \frac{1}{17}, \frac{1}{19}, \frac{1}{21}, \frac{1}{23}, \frac{1}{25}, \frac{1}{27}, \frac{1}{29}\right\},\right.$$
$$\left. 1.83587\right\}$$

We can now see how the partial sums of the rearranged series converge to 2.

```
In[·]:= ListLinePlot[Accumulate[First[Rearrangement[2,
        listpos[2500], listneg[2500]]]]]
```

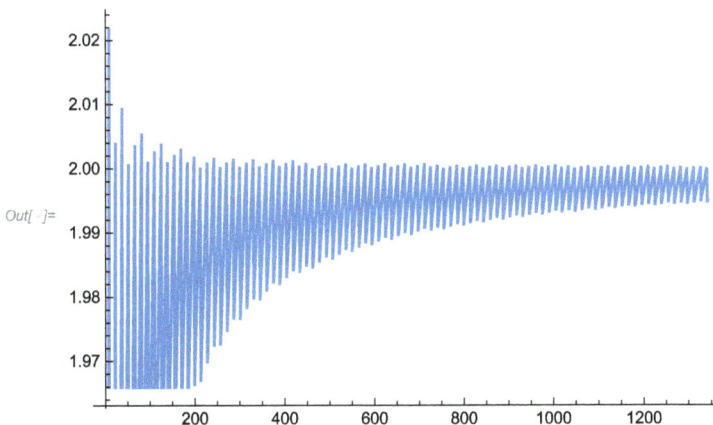

Figure 3.2

3.8 The Cauchy product of series

Suppose we have two infinite series $\sum_{i=0}^{\infty} a_i$ and $\sum_{i=0}^{\infty} b_i$. We know that we can multiply a series by a number (just multiply each term) and we can formally add them:

$$\sum_{i=0}^{\infty} a_i + \sum_{i=0}^{\infty} b_i = \sum_{i=0}^{\infty} (a_i + b_i).$$

We know that this equation is also valid for the sums of the series, whenever both sides are defined.

We would naturally like to define the product of two series in such a way that the analogous property holds, that is, the sum of the product series $\sum_{i=0}^{\infty} a_i \times \sum_{i=0}^{\infty} b_i$ should be the product of the sums of $\sum_{i=0}^{\infty} a_i$ and $\sum_{i=0}^{\infty} b_i$, when again everything is defined. It is obvious that the product must be a sum of the terms $a_i b_j$ for all i and j but the order in which we add them is important (unless we have an absolutely convergent series or a finite sum) for the product series to have the desired property. When we multiply and expand finite sums, Mathematica® arranges the terms in its own standard order but if we used this order for multiplying infinite series, the product series would not have good properties. Therefore, we define the so called Cauchy product of series by "going along the diagonals" in the infinite matrix as in the picture below.

In[·]:= g1 = Graphics[Flatten[Table[If[i - j <= 4,
      Point[{i/3, j/3}], Point[{}]], {i, 0, 4},
      {j, 0, -4, -1}], 1]];

In[·]:= g2 = Graphics[Flatten[Table[If[i - j <= 3,
      Text[Subscript[a, i]* Subscript[b, -j],
      {i/3, j/3}], Text[""], {i/3, j/3}]], {i, 0, 3},
      {j, 0, -3, -1}], 1]];

In[·]:= g3 = Graphics[Table[Line[Table[{(n - i)/3, -i/3},
      {i, 0, n}]], {n, 0, 3}]];

In[·]:= Show[g1, g2, g3]

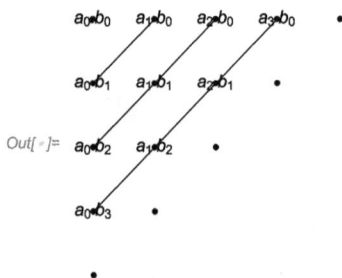

Figure 3.3

Formally, we define the *Cauchy product* by

$$\sum_{i=0}^{\infty} a_i \times \sum_{i=0}^{\infty} b_i = \sum_{n=0}^{\infty} \left(\sum_{i=0}^{n} a_i b_{n-i} \right)$$

$$= a_0 b_0 + (a_0 b_1 + a_1 b_0) + (a_0 b_2 + a_1 b_1 + a_2 b_0) + \cdots.$$

The theorem due to Mertens says that if both series $\sum_{i=0}^{\infty} a_i$ and $\sum_{i=0}^{\infty} b_i$ are convergent and if one of them is absolutely convergent, then the sum of their Cauchy product is the product of their sums.

3.9 Divergent series

We know that a series can be divergent in different ways, it can be divergent to ∞ or $-\infty$ or have no limit. The fact that a series is divergent to infinity can sometimes be useful. For example:

In[·]:= SumConvergence[1/Prime[n], n]
Out[·]:= False

This tells us that the sum of the series $\sum_{n=1}^{\infty} 1/p_n$ is ∞, which implies that there are infinitely many primes. On the other hand,

In[·]:= SumConvergence[(-1)^n, n]
Out[·]:= False

This alone does not tell us anything useful about this series.

We will now see that in many cases we can find a different notion of "sum" of series, which gives a finite answer in the case of some divergent series. This is referred to as "regularized sum" and can be calculated using the option Regularization in Mathematica®'s function Sum. For example, we have

In[·]:= Sum[(-1)^n, {n, 0, Infinity}, Regularization -> "Cesaro"]
Out[·]:= $\dfrac{1}{2}$

This answer looks strange but surprisingly regularization is useful and not only in mathematics (where it is used to prove formulas about sums of convergent series [10, Section 1.2], in solving differential equations [2], etc.) but also in modern physics.

We begin with the notion of regularized sum of a series [10]. In this section we shall use the notation $\sum_{n=1}^{\infty} a_n$ for a series (sequence of partial sums of the sequence $\{a_n\}$, and we will write $S(\sum_{n=1}^{\infty} a_n)$ for its sum (limit) if it exists. Then a *regularized sum* is a function $S^* : \mathcal{S} \to \mathbb{R}$ (more generally $\mathbb{C}$), where $\mathcal{S}$ is a set of all series, with the following properties:

1. $S^*(\sum_{n=1}^{\infty} a_n) = S(\sum_{n=1}^{\infty} a_n)$ for every convergent series $\sum_{n=1}^{\infty} a_n$.

2. $S^*(a_0 + \sum_{n=1}^{\infty} a_n) = a_0 + S^*(\sum_{n=1}^{\infty} a_n)$ (invariance with respect to addition).
3. $S^*(\sum_{i=1}^{\infty}(a_i + b_i)) = S^*(\sum_{i=1}^{\infty} a_i) + S^*(\sum_{i=1}^{\infty} b_i)$.
4. $S^*(\sum_{i=1}^{\infty} \lambda a_i) = \lambda S^*(\sum_{i=1}^{\infty} a_i)$.
5. $S^*(\sum_{i=1}^{\infty} a_i \times \sum_{i=1}^{\infty} b_i) = S^*(\sum_{i=1}^{\infty} a_i) S^*(\sum_{i=1}^{\infty} b_i)$, where $\times$ is the Cauchy product.

Consider the divergent series $\sum_{n=0}^{\infty}(-1)^n$. Let us assume that there is some S^* which satisfies properties 1–5 above and such that $S^*(\sum_{n=0}^{\infty}(-1)^n) = S_0 \in \mathbb{R}$. Then it is easy to find S_0. Indeed,

$$S_0 = S^*\left(\sum_{n=0}^{\infty}(-1)^n\right) = S^*\left(1 + \sum_{n=1}^{\infty}(-1)^n\right) = 1 + S^*\left(\sum_{n=1}^{\infty}(-1)^n\right) = 1 - S_0.$$

Hence, $S_0 = 1/2$. Of course, this does mean that such regularization exists and clearly it should be given explicitly.

To define a regularization one needs to find a function S^* with the above properties. For example, the Cesàro regularization is based on the following theorem (see, for instance, [14, Theorem 2.64]).

Theorem 8. *Let a_n be a sequence with limit $g \in \mathbb{R} \cup \{\infty, -\infty\}$. Then the sequence of arithmetic means $\{(\sum_{i=1}^{n} a_i)/n\}$ has limit g.*

In Mathematica® we can create means from lists by using for instance

In[·]:= (#1/Length[#1] &) /@ Rest[Accumulate[{a, b, c, d, e}]]
Out[·]:= $\{\frac{a+b}{2}, \frac{1}{3}(a+b+c), \frac{1}{4}(a+b+c+d), \frac{1}{5}(a+b+c+d+e)\}$

Note that the sequence of means of a sequence may be convergent when the sequence itself has no limit. The standard example is $\{(-1)^n\}$, which does not have a limit, while

In[·]:= DiscreteLimit[Sum[(-1)^k, {k, 1, n}]/n, n -> Infinity]
Out[·]:= 0

Theorem 8 is the basis for the *Cesàro regularization*, in which we define the Cesàro sum of a series as the limit of the sequence of means of the sequence of partial sums of the sequence. That is,

$$S^*\left(\sum_{i=1}^{\infty} a_i\right) = \lim_{m \to \infty} \frac{\sum_{n=0}^{m} \sum_{k=0}^{n} a_k}{m+1}. \tag{3.8}$$

It is easy to check that S^* defined in (3.8) satisfies conditions 1–5 above.

An important application of Cesàro's regularization is Cesàro's theorem on the Cauchy product of two convergent series. The theorem states that the Cauchy product is Cesàro summable and its Cesàro sum is the product of the sums of two series.

There are other implemented regularization methods in Mathematica® (the Abel method, which uses power series discussed in Section 3.10, the Borel method, which uses integration discussed later on as well, the Dirichlet method, which uses a function series, and the Euler methods for alternating sums).

We will not consider any applications of divergent series here except for remarking that with their help one can prove various identities involving convergent series (see examples in [10]).

For example, let t be a number such that $-\pi \le t \le \pi$. Then the following identity holds:

$$\sum_{n=1}^{\infty} \frac{(-1)^{n-1}(1-\cos(tn))}{n^2} = \frac{t^2}{4}. \tag{3.9}$$

The proof given in [10] involves expanding the series on the left (which we can prove is convergent) in terms of t and then rearranging and expressing it in terms of divergent series which can be summed using a regularization method. Because the original series is convergent, the answer that one gets using this method must be equal to the ordinary sum of the series.

Mathematica® gives a much more complicated answer for the left hand side, namely,

In[·]:= FullSimplify[Sum[((-1)^(n - 1)*(1 - Cos[t*n]))/n^2,
       {n, 1, Infinity}], Assumptions -> -Pi <= t <= Pi]

Out[·]:= $\frac{1}{12} \left(\pi^2 + 6\,\mathrm{PolyLog}\left[2, -E^{-I\,t}\right] + 6\,\mathrm{PolyLog}[2, -E^{I\,t}]\right)$

and Mathematica® is unable to prove the above identity. In situations, when we have an identity we believe to be true but which Mathematica® cannot prove, it is still possible to provide some evidence for the truth (or possibly disproving it) by substituting random numbers lying within some range. However, using MachinePrecision random numbers may sometimes not be sufficient to confirm such an identity, because the inaccuracy of calculations may cause the program to give different answers on both sides. This is a situation in which Mathematica®'s ability to control precision can be useful.

In[·]:= Sum[((-1)^(n - 1)*(1 - Cos[t*n]))/n^2,
       {n, 1, Infinity}] == t^2/4 /. Transpose[{Thread[t ->
       RandomReal[{-Pi, Pi}, {7}, WorkingPrecision -> 10]]}]
Out[·]:= {True, True, True, True, True, True, True}

Using only MachinePrecision we sometimes can get False:

In[·]:= Sum[((-1)^(n - 1)*(1 - Cos[t*n]))/n^2,
       {n, 1, Infinity}] == t^2/4 /.
       Transpose[{Thread[t -> RandomReal[{-Pi, Pi}, {7}]]}]
Out[·]:= {False, True, True, True, True, True, True}

3.10 Power series

A *power series* is a series of the form $\sum_{n=0}^{\infty} a_n(x - x_0)^n$, where $\{a_n\}$ is a sequence of real (or complex) numbers called the coefficients of the power series and x is in $\mathbb{R}$

or $\mathbb{C}$. The number x_0 is called the *center of the power series*. The set of points $\{x :$ $\sum_{n=0}^{\infty} a_n(x - x_0)^n$ converges$\}$ is called the *region of convergence of the series*. It is always non-empty because it always contains the point x_0. One can show that the region of con-vergence is always an interval which could be just the point x_0, a finite interval (closed, open or half open) or the entire real line (in the complex case the region of convergence is a disk). There are formulas to find the radius of convergence of power series [14] which can be derived from convergence tests. Note that the radius of convergence does not de-termine the region of convergence and the endpoints of the interval have to be checked separately.

The function SumConvergence can also be used to find the region of convergence of power series. Here are a few examples.

In[·]:= SumConvergence[n!*x^n, n]
Out[·]:= x == 0

The region of convergence of $\sum_{n=1}^{\infty} n! x^n$ is just the center of the series. This is easily proved by using either the root or the ratio test.

In[·]:= SumConvergence[(x - 1)^n/n, n]
Out[·]:= Abs[-1 + x] <= 1 && x ≠ 2

In the real case this says that the series $\sum_{n=1}^{\infty}(x-1)^n/n$ converges on the half open interval $[0, 2)$. Convergence inside the open interval $(0, 2)$ can again be easily proved by using the root or ratio test, but the endpoints 0 and 2 must be considered separately. For $x = 0$ we get $\sum_{n=1}^{\infty}(-1)^n/n$ which we know is convergent (by the Dirichlet test) and for $x = 2$ we get the harmonic series, which we know is divergent. In the complex case the result says that the series is convergent on the entire closed disk $|x - 1| \le 1$ except at $x = 2$.

Our next example is the following:

In[·]:= SumConvergence[x^n/n!, n]
Out[·]:= True

This says that the series $\sum_{n=1}^{\infty} x^n/n!$ converges on the entire real line (complex plane).

Let $\sum_{n=0}^{\infty} a_n(x - x_0)^n$ be a series with a region of convergence $I \subset \mathbb{R}$. Then $f(x) :=$ $\sum_{n=0}^{\infty} a_n(x - x_0)^n$ defines a function $f : I \to \mathbb{R}$. Functions defined in this way are called *analytic functions* and we will see that they possess many properties of polynomials. Here are some important analytic functions:

$$\exp(x) = \sum_{n=0}^{\infty} \frac{x^n}{n!},$$

$$\sin(x) = \sum_{n=0}^{\infty} \frac{(-1)^n}{(2n + 1)!} x^{2n+1},$$

$$\cos(x) = \sum_{n=0}^{\infty} \frac{(-1)^n}{(2n)!} x^{2n}.$$

All of these series have the region of convergence $\mathbb{R}$ (or $\mathbb{C}$ in the complex plane), hence the corresponding functions are defined everywhere. Such analytic functions are called *entire*. These functions are equal to the familiar functions usually defined in a different way in elementary calculus courses. One can show that the functions cos and sin are indeed equal to the usual trigonometric functions defined in terms of ratios of sides of right angled triangles and $\exp(x) = e^x$, where e was defined earlier as a limit of a certain sequence.

In Mathematica® one can check the analyticity of both real and complex functions:

In[·]:= FunctionAnalytic[Sin[x], x]
Out[·]:= True

In[·]:= FunctionAnalytic[E^z, z, Complexes]
Out[·]:= True

In[·]:= FunctionAnalytic[{Log[z], -Pi < Arg[z] < Pi
            && z != 0}, z]
Out[·]:= True

The function Asymptotic can give power series representation for functions, for example

In[·]:= Asymptotic[1/(1 - z), {z, 0, Infinity}]
Out[·]:= $\sum_{n=0}^{\infty} z^n$

It can also compute the radius of convergence of the power series:

In[·]:= Asymptotic[1/(1 - z), {z, 0, Infinity},
            GenerateConditions -> True]
Out[·]:= $\sum_{n=0}^{\infty} z^n$ if Abs[z] < 1

In[·]:= Asymptotic[Sin[z], {z, 0, Infinity},
            GenerateConditions -> True]
Out[·]:= $\sum_{n=0}^{\infty} \frac{(-1)^n z^{1+2n}}{(1+2n)!}$

When Mathematica® cannot find an expression for the general term of a series, it returns the input:

In[·]:= Asymptotic[1/Sin[z], {z, 0, Infinity}]
Out[·]:= Csc[z]

Mathematica®, of course, can compute any finite number of terms of the series:

In[·]:= Asymptotic[1/Sin[z], {z, 0, 5}]
Out[·]:= 1/z + z/6 + (7*z^3)/360 + (31*z^5)/15120

We can use the properties of the Cauchy product of series to prove that these functions defined by power series have the expected properties. For example, let us show the identity

$$\exp(x + y) = \exp(x)\exp(y).$$

We will prove this by showing that the same identity holds when exp denotes the series itself, rather than its sum and the product on the right denotes the Cauchy product of the series. Then from Mertens' theorem it will follow that our identity (in which exp means the sum of the series) is true.

The expression on the left is the sum of a power series in $z = x + y$, which after expansion of each term $(x + y)^n$ becomes a power series in two variables x and y. We can find the coefficient of $x^n y^m$ by using Mathematica®'s function SeriesCoefficient (which will be considered in greater detail later, when we study the Taylor expansion of functions):

In[·]:= SeriesCoefficient[Exp[x + y], {x, 0, n}, {y, 0, m}]

$$Out[\cdot]:= \begin{cases} \dfrac{1}{m!\,n!} & n \geq 0 \wedge m \geq 0 \\ 0 & \text{True} \end{cases}$$

Note that this kind of expression in Mathematica® the word True should be interpreted as saying "otherwise".

Now we do the same to the power series on the right hand side and we see that the coefficients are the same. This means that this series is the same as the series on the left hand side:

In[·]:= SeriesCoefficient[Exp[x]*Exp[y], {x, 0, n}, {y, 0, m}]

$$Out[\cdot]:= \begin{cases} \dfrac{1}{m!\,n!} & n \geq 0 \wedge m \geq 0 \\ 0 & \text{True} \end{cases}$$

In some cases Mathematica® may not give the same result unless we use FullSimplify possibly with certain assumptions. Consider for instance the following trigonometric identity:

In[·]:= TrigExpand[Cos[x + y]]
Out[·]:= Cos[x] Cos[y] - Sin[x] Sin[y]

If we use the same method as above, Mathematica® will not return the same expressions on the right and the left hand sides. However, we can show that they are equivalent by using FullSimplify with suitable assumptions:

In[·]:= FullSimplify[SeriesCoefficient[Cos[x + y], {x, 0, n},
        {y, 0, m}], Assumptions -> Element[{m, n}, Integers]
        && m >= 0 && n >= 0]

$$Out[\cdot]:= \frac{\cos\left[\frac{1}{2}(m + n)\pi\right]}{m!\,n!}$$

In[·]:= FullSimplify[SeriesCoefficient[Cos[x]*Cos[y] -
        Sin[x]*Sin[y], {x, 0, n}, {y, 0, m}], Assumptions ->
        Element[{m, n}, Integers] && m >= 0 && n >= 0]

$$Out[\cdot]:= \frac{\cos\left[\frac{1}{2}(m + n)\pi\right]}{m!\,n!}$$

In general, we cannot expect that applying FullSimplify to two different but equal expressions will produce the same expression. For this reason it is usually better to apply FullSimplify to the difference of the expressions and check whether Mathematica® returns 0:

In[·]:= FullSimplify[SeriesCoefficient[Cos[x + y], {x, 0, n},
  {y, 0, m}] - SeriesCoefficient[Cos[x]*Cos[y] -
  Sin[x]*Sin[y], {x, 0, n}, {y, 0, m}], Assumptions ->
  Element[{m, n}, Integers] && m >= 0 && n >= 0]

Out[·]:= 0

In certain problems in mathematics one needs to compute the Padé approximant[1] of a function. Mathematica®'s function PadeApproximant can be used for this purpose. For example, the order [2/3] Padé approximant for the exponential function is given by

In[·]:= PadeApproximant[Exp[x], {x, 0, {2, 3}}]

Out[·]:= $\dfrac{\frac{x^2}{20} + \frac{2x}{5} + 1}{-\frac{x^3}{60} + \frac{3x^2}{20} - \frac{3x}{5} + 1}$

The first $m + n$ terms of the series expansion of the approximant agrees with the ordinary series of the function by using Mathematica®'s function Series (which will be considered in detail when we study the Taylor expansion of functions):

In[·]:= Normal[Series[%, {x, 0, 3}]]

Out[·]:= $1 + x + \frac{x^2}{2} + \frac{x^3}{6}$

In[·]:= Normal[Series[Exp[x], {x, 0, 3}]]

Out[·]:= $1 + x + \frac{x^2}{2} + \frac{x^3}{6}$

For $n = 0$ PadeApproximant gives m terms of the ordinary series:

In[·]:= PadeApproximant[Exp[x], {x, 0, {3, 0}}]

Out[·]:= $1 + x + \frac{x^2}{2} + \frac{x^3}{6}$

Another problem that often occurs in applied mathematics is finding the generating function from its n-th series coefficient. For example, the generating function for the sequence whose n-th term is 1:

In[·]:= GeneratingFunction[1, n, x]

Out[·]:= $\frac{1}{1 - x}$

All the coefficients of the series are 1:

In[·]:= Normal[Series[%, {x, 0, 4}]]

Out[·]:= $1 + x + x^2 + x^3 + x^4$

or

---

1 https://en.wikipedia.org/wiki/Pad%C3%A9_approximant

In[·]:= Asymptotic[%%, {x, 0, Infinity}]
Out[·]:= $\sum_{n=0}^{\infty} x^n$

In[·]:= GeneratingFunction[1/n!, n, x]
Out[·]:= E^x

In[·]:= GeneratingFunction[a^n, n, x, GenerateConditions -> True]
Out[·]:= $\frac{1}{1-ax}$ if Abs[x] < 1/Abs[a]]

One can also try to find a generating function for a sequence represented by a finite list:

In[·]:= FindGeneratingFunction[{1, 1, 2, 3, 5, 8, 13}, x]
Out[·]:= $\frac{1}{1-x-x^2}$

In[·]:= Normal[Series[%, {x, 0, 7}]]
Out[·]:= $1 + x + 2x^2 + 3x^3 + 5x^4 + 8x^5 + 13x^6 + 21x^7$

As always, when Mathematica® cannot find a generating function, it returns the input:

In[·]:= FindGeneratingFunction[{1, 1, 2, 3, 5, 8, 13, 15}, x]
Out[·]:= FindGeneratingFunction[{1, 1, 2, 3, 5, 8, 13, 15}, x]

4 Limits of functions and continuity

In this chapter we will extend the concept of limit to arbitrary real-valued functions of one real variable, define continuity and study some properties of continuous functions and related topics. We state without proof several important theorems on continuous functions and give some examples of their applications.

4.1 Limits of functions

Let $A \subset \mathbb{R}$. We say that a point $x \in \mathbb{R} \cup \{\infty, -\infty\}$ is a *limit point* of A if there exists a sequence $\{a_n\}$ with $a_n \in A$, $a_n \neq x$ for all n such that $\lim_{n \to \infty} a_n = x$. Note that x itself need not be in A. In fact ∞ is the only limit point of $\mathbb{N}$. Another example is an open interval $A = (a, b)$, in which case a and b are both limit points although they do not belong to A. Of course, a point of A can be (but need not be) a limit point. The most common type of limit point that belongs to A is an interior point, that is, a point that is contained in an open interval which is a subset of A.

Let x_0 be a limit point of A and let $f : A \to \mathbb{R}$ be a function. We say that $g \in \mathbb{R} \cup \{\infty, -\infty\}$ is the *limit of function* f as $x \to x_0$ (written as $\lim_{x \to x_0} f(x) = g$) if for every sequence $x_n \to x_0$ we have

In[·]:= DiscreteLimit[f[x[n]], n -> Infinity] == g

Out[·]:= $\lim_{n \to \infty_{\mathbb{Z}}} f[x[n]]$ == g

This definition illustrates clearly the distinction between Mathematica®'s functions DiscreteLimit and Limit. Limit is defined for functions in terms of DiscreteLimit, which is defined for sequences.

This definition of limit is known as *Heine's definition*. It is equivalent to another definition known as *Cauchy's definition*. Unlike Heine's definition, Cauchy's definition takes a different form in the cases when both x_0 and g are real numbers and in the cases when any of them is either ∞ or $-\infty$. This means that there are nine different definitions, though as they are very similar we will only state two of them.

Definition 1. Let $a \in \mathbb{R}$ be a limit point of $A \subset \mathbb{R}$, let $f : A \to \mathbb{R}$ be a function and let $g \in \mathbb{R}$. Then $\lim_{x \to a} f(x) = g$ if and only if for every $\varepsilon > 0$ there is a $\delta > 0$ such that $0 < |x - a| < \delta$ implies $|f(x) - g| < \varepsilon$.

Note that δ depends on both a and ε.

Definition 2. Suppose that ∞ is a limit point of A (e. g., $A = \mathbb{R}$). Then $\lim_{x \to \infty} f(x) = \infty$ if for every $M > 0$ there exists $N > 0$ such that $x > N$ implies $f(x) > M$.

The function Limit returns conditional answers and admits assumptions:

In[·]:= Limit[x^a, x -> Infinity]

Out[·]:= ConditionalExpression[∞, a > 0]

https://doi.org/10.1515/9783111533063-004

In[·]:= `Limit[x^a, x -> Infinity, Assumptions -> {a < 0}]`
Out[·]:= 0

If for some reason we wish to get an unconditional answer we can use the option `GenerateConditions`:

In[·]:= `Limit[x^a, x -> Infinity, GenerateConditions -> False]`
Out[·]:= ∞

We should do this only in case we are sure that $a > 0$.

Limits of functions at a point have properties analogous to those of limits of sequences and their proofs reduce to the proofs of analogous results for sequences. For example, it is easy to extend statements about the limits of sums, products and reciprocals of sequences to sums, products and reciprocals of functions. We will thus omit these statements (see [14, Chapter 3]).

Here are some examples of computation of certain important limits which can be proved by using the Taylor series:

In[·]:= `Limit[(Exp[x] - 1)/x, x -> 0]`
Out[·]:= 1

In[·]:= `Limit[Sin[x]/x, x -> 0]`
Out[·]:= 1

In[·]:= `Limit[(Cos[x] - 1)/x^2, x -> 0]`
Out[·]:= $-\dfrac{1}{2}$

4.2 One-sided limits

Let $f : A \to \mathbb{R}$ be a function and $a \in \mathbb{R}$ be a limit point. We say that g is a *limit of f at a from the right (or from above)* if for each sequence $x_n \to a$ such that $x_n > a$ for all $n \in \mathbb{N}$ we have $\lim_{n \to \infty} f(x_n) = g$. In this case we write $\lim_{x \to a^+} f(x) = g$. We say that g is a *limit of f at a from the left (or from below)* if for each sequence $x_n \to a$ such that $x_n < a$ for all $n \in \mathbb{N}$ we have $\lim_{n \to \infty} f(x_n) = g$. We write $\lim_{x \to a^-} f(x) = g$. We have the following theorem.

Theorem 9. *The limit $\lim_{x \to a} f(x) = g$ if and only if $\lim_{x \to a^-} f(x)$ and $\lim_{x \to a^-} f(x)$ both exist and are equal to g.*

Of course in the case when $a = \infty$ all limits are actually limits from the left and when $a = -\infty$ all limits are limits from the right.

As we already know, the function `Limit` computes limits of functions in Mathematica®. It provides one of the cases in which a significant change took place in Mathematica® in version 11, which does not preserve "backward compatibility" (i. e., certain outputs will be different when evaluated in earlier versions of Mathematica®). The main difference is that before version 11 Mathematica®'s `Limit` by default always computed the limit from the right:

In[·]:= `Limit[1/x, x -> 0]`
Out[·]:= ∞

To get the limit from the left one had to use the options `Direction`:

In[·]:= `Limit[1/x, x -> 0, Direction -> 1]`
Out[·]:= $-\infty$

The two limits are not equal, hence the two-sided limit does not exist. This is indeed the answer that `Limit` returns in Mathematica® 11:

In[·]:= `Limit[1/x, x -> 0]`
Out[·]:= `Indeterminate`

This, as we already know, means that the limit does not exist.

To compute one-sided limits in Mathematica® 11, we have to use the option `Direction`, which still can take the values 1 and –1 as in older versions or "FromAbove" and "FromBelow":

In[·]:= `Limit[1/x, x -> 0, Direction -> "FromAbove"]`
Out[·]:= ∞

One can also compute the limit in the complex plane, for example,

In[·]:= `Limit[1/x, x -> 0, Direction -> "Complexes"]`
Out[·]:= `ComplexInfinity`

Here we should mention another difference in the behavior of the function `Limit` in Mathematica® 11 and in the earlier versions. Consider the limit of the function $\sin(1/x)$ as x tends to 0. Clearly this limit does not exist:

In[·]:= `Plot[Sin[1/x], {x, -1, 1}, Exclusions -> {x == 0}]`

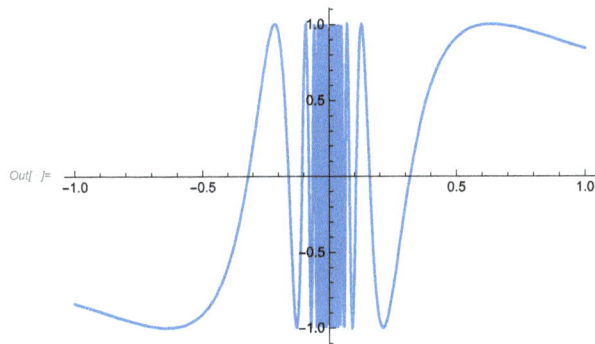

Figure 4.1

In Mathematica® 10 we obtain

In[·]:= `Limit[Sin[1/x], x -> 0, Direction -> -1]`
Out[·]:= `Interval[{-1,1}]`

This is a non-standard answer which reflects the fact that as *x* approaches 0 (even from above) the values of the function assume all values in the interval $[-1,1]$. In Mathematica® 11, however, we obtain

In[·]:= Limit[Sin[1/x], x -> 0, Direction -> "FromAbove"]
Out[·]:= Indeterminate

which agrees with the usual convention that this kind of limit does not exist.

In recent versions of Mathematica® it is possible to use infinite limits in some plotting functions. This often allows us to see the global behavior of a function (however, the scaling is also sometimes applied, as seen from the following graphs).

In[·]:= Plot[Sin[x]/x, {x, 1, Infinity}, AxesOrigin ->
    {0, 0}]

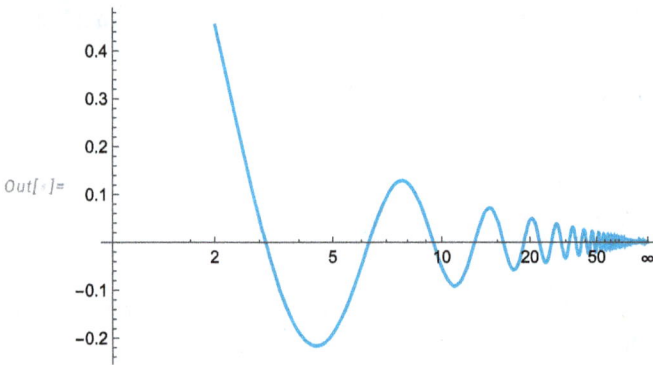

Figure 4.2

In[·]:= Plot[Sin[x], {x, -Infinity, Infinity}]

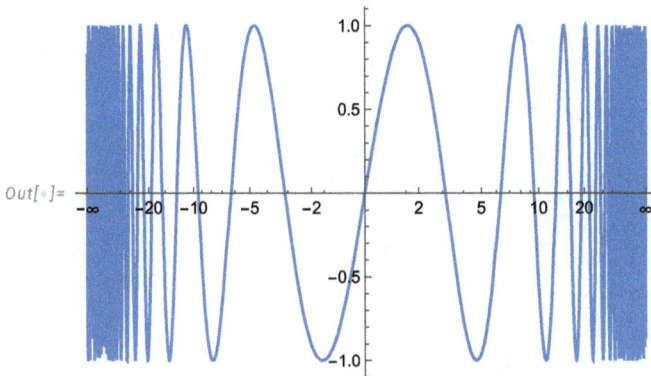

Figure 4.3

Sometimes Mathematica® draws the plot at infinity incorrectly, as can be seen from the following example below. One can easily check that

In[·]:= `Limit[Log[x^2007 + x^2009]/x^(1/9999), x -> Infinity]`
Out[·]:= 0

but the plot suggests something different, namely that the function tends to infinity as x tends to infinity:

In[·]:= `Plot[Log[x^2007 + x^2009]/x^(1/9999), {x, 500, Infinity}]`

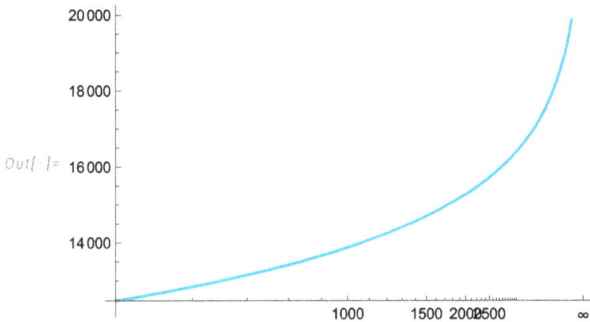

Figure 4.4

Adding `WorkingPrecision` does not help as this yields a lot of warnings (which we omit):

In[·]:= `Plot[Log[x^2007 + x^2009]/x^(1/9999), {x, 500,`
        `Infinity}, WorkingPrecision -> 100]`

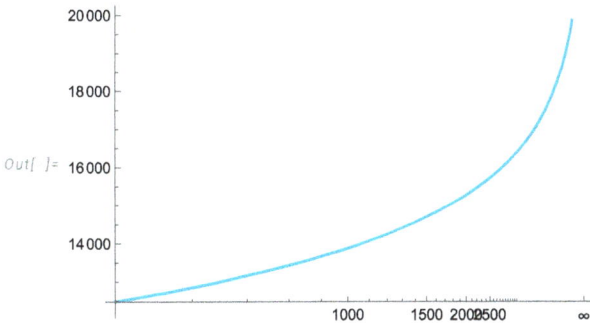

Figure 4.5

An attempt to draw the graph of the function over large but finite interval also fails (we again omit the warning):

In[·]:= `Plot[Log[x^2007 + x^2009]/x^(1/9999),`
        `{x, 10^1000, 10^1020}]`
Out[·]:= `Plot[Log[x^2007 + x^2009]/x^(1/9999),`
        `{x, 10^1000, 10^1020}]`

This is due to the restriction on the size of numbers which can be used in plots. One can check that the value of the function for $x = 10^{10000}$

In[·]:= N[Log[x^2007 + x^2009]/x^(1/9999) /.
      x -> 10^10000, 20]
Out[·]:= 4.6248283166140510150*10^6

is very large. Only after substituting $x = 10^{100000}$ one can see that the value of the function is close to zero:

In[·]:= N[Log[x^2007 + x^2009]/x^(1/9999) /.
      x -> 10^100000, 20]
Out[·]:= 0.046152531293144629346

Mathematica® cannot use numbers greater that

In[·]:= $MaxMachineNumber
Out[·]:= 1.79769*10^308

in plots. It seems clear that Plot does not use analytic methods to determine the behavior of functions at infinity but relies on numerical computations. Note that the biggest number in Mathematica® is

In[·]:= $MaxNumber
Out[·]:= 1.605216761933662*10^1355718576299609

4.3 Continuous functions

Definition 3. Let $a \in A \subset \mathbb{R}$ and let $f : A \to \mathbb{R}$ be a function. We say that f *is continuous at a* if for every sequence $a_n \to a$ we have $\lim_{n\to\infty} f(a_n) = f(\lim_{n\to\infty} a_n) = f(a)$. If f is continuous at all points in its domain we say that it is *continuous*. There is also a natural notion of a left and right continuity at a point a. If a function is not continuous at a, then we say that it is *discontinuous at a*.

Note that a does not have to be a limit point of A since every function is continuous at an isolated point (because in such a case the only sequences convergent to the point are the constant sequences).

Note also that we do not consider continuity of a function at a point which does not lie in its domain. Thus the function $f : \mathbb{R} \setminus \{0\} \to \mathbb{R}$ which takes x to $1/x$ is continuous everywhere. However, it cannot be extended to a continuous function $f : \mathbb{R} \to \mathbb{R}$. Any such extended function will be discontinuous at 0. For instance if we extend the function $f(x) = 1/x$ with $f(0) = 3$, then it is obviously discontinuous:

In[·]:= Plot[1/x, {x, -2, 2}, Exclusions -> 0,
      Epilog -> Point[{0, 3}]]

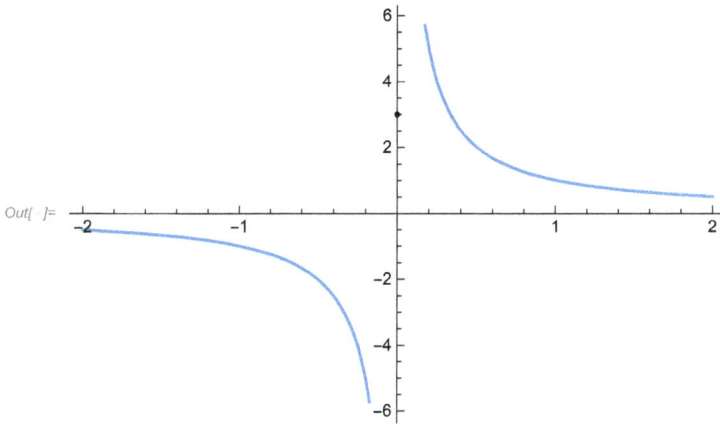

Figure 4.6

The above definition of continuity is given in Heine's form. There is also Cauchy's version, which we state as a theorem.

Theorem 10. *A function $f : A \to \mathbb{R}$ is continuous at $a \in A$ if and only if for every $\varepsilon > 0$ there exists $\delta > 0$ such that $|f(x) - f(a)| < \varepsilon$ whenever $x \in A$ and $|x - a| < \delta$.*

An interesting consequence of this theorem is that we can prove the continuity of certain functions by using quantifier elimination. For example, we can prove that the function $f(x) = x^2 - x + 1$ is continuous everywhere:

```
In[·]:= f[x_] := x^2 - x + 1
In[·]:= Resolve[ForAll[{e, x}, e > 0, Exists[d, d > 0,
          ForAll[{y}, Element[{y}, Reals] && Abs[x - y] < d,
          Abs[f[x] - f[y]] < e]]]]
Out[·]:= True
```

Of course, this is not a practical approach to proving continuity of functions. Instead, one starts by proving that the set of continuous functions (with a fixed domain) is closed under the operations of addition, multiplication, division (when defined) and composition of function, i. e., that the set of continuous functions has the structure of an algebra (see [14]). Since it is obvious that all constant functions are continuous and that the identity function $x \mapsto x$ is continuous, it follows that all rational functions are continuous.

Another important set of continuous functions is the set of the *analytic functions*, i. e., functions defined by a power series. As we stated in Chapter 3, such a series is always convergent in an interval, which could be open, closed, half open and bounded or unbounded. One can show that a function defined in this way is always continuous in its entire region of convergence (if the interval has endpoints, the function is left or right continuous at these endpoints). This is Abel's Limit Theorem (see [14, p. 416]).

Mathematica® can determine many properties of a function given by a formula. For example:

```
In[·]:= FunctionDomain[x/(x^4 - 1), x]
Out[·]:= x < -1 || -1 < x < 1 || x > 1
```

```
In[·]:= FunctionRange[E^x, x, y]
Out[·]:= y > 0
```

```
In[·]:= FunctionInjective[E^x, x]
Out[·]:= True
```

```
In[·]:= FunctionSurjective[Tan[x], x]
Out[·]:= True
```

```
In[·]:= FunctionBijective[Sinh[x], x]
Out[·]:= True
```

```
In[·]:= FunctionPeriod[Sin[x], x]
Out[·]:= 2 Pi
```

See also guide/FunctionProperties.

Various aspects of continuity can also be determined:

```
In[·]:= FunctionContinuous[Sin[x], x]
Out[·]:= True
```

```
In[·]:= FunctionContinuous[{Log[x], x > 0}, x]
Out[·]:= True
```

```
In[·]:= FunctionContinuous[Cos[x]/a, x,
          GenerateConditions -> Automatic]
Out[·]:= ConditionalExpression[True, Element[a, Reals]]
```

Continuity of complex functions can be checked similarly:

```
In[·]:= FunctionContinuous[Sqrt[x], x, Complexes]
Out[·]:= False
```

For continuous functions, limits can be computed by substitution:

```
In[·]:= {FunctionContinuous[Cosh[x], x],
          Limit[Cosh[x], x -> 0], Cosh[0]}
Out[·]:= {True, 1, 1}
```

Mathematica® can also find the discontinuities of some functions, for example,

```
In[·]:= FunctionDiscontinuities[Tan[x], x]
Out[·]:= Cos[x] == 0
```

Note that Mathematica® considers a function to be discontinuous at points where no value is defined. This does not agree with our definition according to which function can only be continuous or discontinuous at points where it has a defined value.

4.4 Discontinuous functions

Let us now consider some examples of *discontinuous functions*. The simplest type of discontinuity is a simple jump at a point. In one such situation both one-sided limits exist and are equal but are not equal to the value of the function at the point. Such a function can be made continuous by changing its value at a single point. The other kind of jump discontinuity is when the function is either left continuous or right continuous but not continuous as in the picture below. The best way to produce functions with this property in Mathematica® is by using the function Piecewise, e. g.,

In[·]:= Plot[Piecewise[{{x, x <= 1}, {1 + x^2, x > 1}}],
        {x, -2, 2}]

Out[ ]=

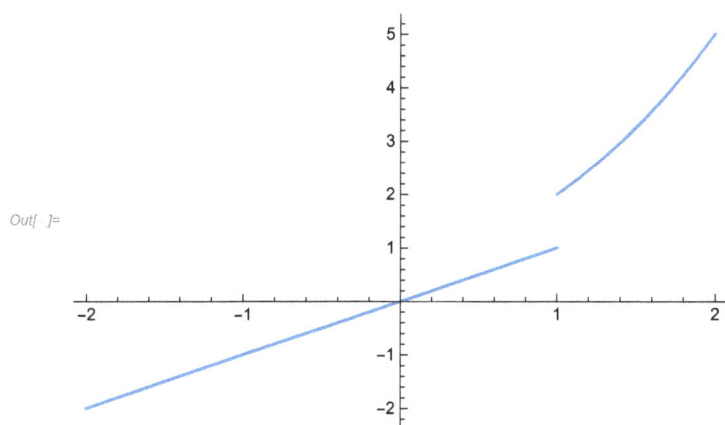

Figure 4.7

This function is obtained by joining two continuous functions whose right and left limits at 1 do not agree. Indeed, we can compute the limits

In[·]:= Limit[Piecewise[{{x, x <= 1}, {1 + x^2, x > 1}}],
        x -> 1, Direction -> "FromAbove"]
Out[·]:= 2

In[·]:= Limit[Piecewise[{{x, x <= 1}, {1 + x^2, x > 1}}],
        x -> 1, Direction -> "FromBelow"]
Out[·]:= 1

In[·]:= Limit[Piecewise[{{x, x <= 1}, {1 + x^2, x > 1}}],
  x -> 1]
Out[·]:= Indeterminate

This is an example of a left continuous function.

We could of course use the function If to obtain the same plot as above by using

In[·]:= Plot[If[x <= 1, x, 1 + x^2], {x, -2, 2},
  Exclusions -> 1]

However Mathematica® cannot reliably perform mathematical operations such as differentiation and integration on piecewise functions defined by means of If. For example in the following expression

In[·]:= D[If[x <= 1, x, 1 + x^2], x]
Out[·]:= If[x <= 1, 1, 2 x]

Mathematica® does not notice that the function is not differentiable at 1, whereas using Piecewise it gives the correct answer:

In[·]:= D[Piecewise[{{x, x <= 1}, {1 + x^2, x > 1}}], x]

$$Out[·]:= \begin{cases} 1 & x < 1 \\ 2\,x & x > 1 \\ \text{Indeterminate} & \text{True} \end{cases}$$

Another important kind of discontinuity is when the one-sided limits do not exist. For example

In[·]:= f[x_] := Piecewise[{{Sin[1/x], x != 0}}, 1]

This situation is better seen by using an interactive graphic representation:

In[·]:= Manipulate[Show[Plot[f[x], {x, 1 - d, 1},
    PlotRange -> {{-1, 1}, {-1, 1.2}}], Plot[f[x],
    {x, -1, -1 + d}, PlotRange -> {{-1, 1}, {-1, 1.2}}],
    Epilog -> {PointSize[0.02], Point[{{0, 1}}]}],
    {{d, 0.1, "d"}, 0.01, 0.99, 0.01}]

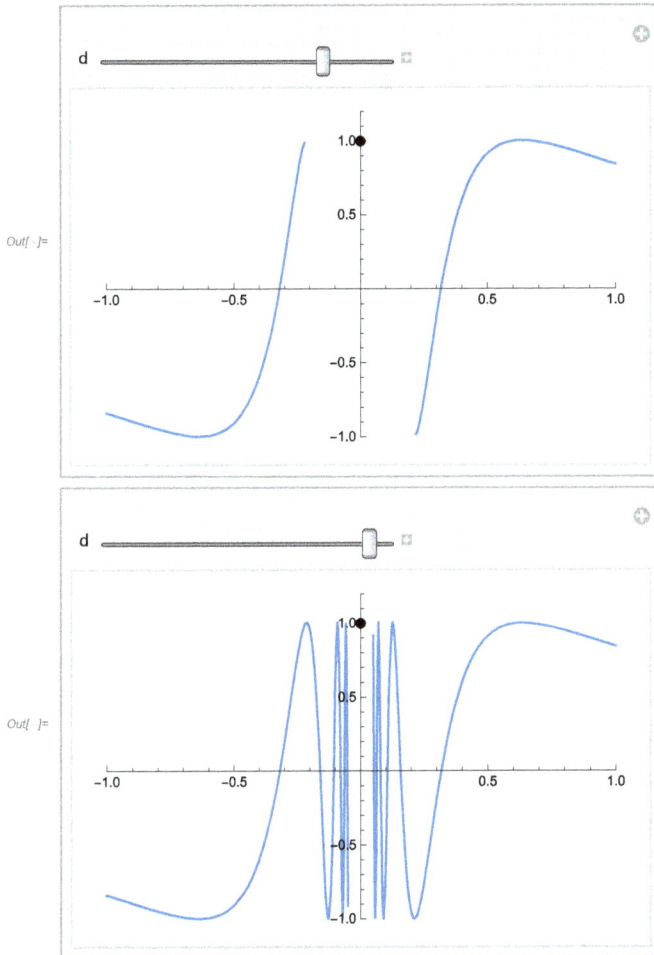

Figure 4.8

In the next example the limit from the left exists but the limit from the right does not. This is another example of the left continuous function.

```
In[·]:= Plot[Piecewise[{{x, x <= 1}, {Sin[1/(x - 1)],
        x > 1}}], {x, -2, 2}]
```

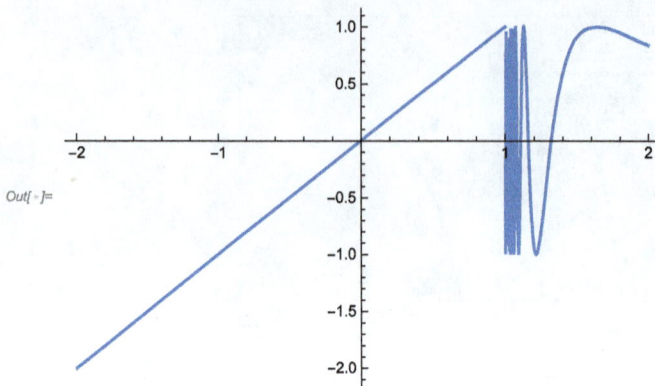

Figure 4.9

We can check that the limit from above does not exist

In[·]:= `Limit[Piecewise[{{x, x <= 1}, {Sin[1/(x - 1)],`
      `x > 1}}], x -> 1, Direction -> "FromAbove"]`
Out[·]:= `Indeterminate`

As we know, many kinds of functions, including discontinuous ones, can be defined by using `Piecewise`. It is possible to expand a complicated nested piecewise expression to a single piecewise function:

In[·]:= `PiecewiseExpand[Piecewise[{{Piecewise[{{1,`
      `x < 1}}, 2], x > 0}}, 3]]`
Out[·]:= `Piecewise[{{1, 0 < x < 1}, {2, x >= 1}}, 3]`

Assumptions can be specified both as an argument and as an option value:

In[·]:= `PiecewiseExpand[Floor[x], x > 0, Assumptions -> x < 3]`
Out[·]:= `Piecewise[{{1, 1 <= x < 2}, {2, x >= 2}}, 0]`

4.4.1 Example: the Dirichlet function

One can define in Mathematica® more complicated functions, for example the famous Dirichlet function, which is not continuous at any point:

In[·]:= `dirichlet[x_] := Boole[Simplify[Element[x, Rationals]]]`

This function is equal to 1 for rational numbers and is equal to 0 otherwise:

In[·]:= `dirichlet /@ {Sqrt[2], 1/2, Pi}`
Out[·]:= `{0, 1, 0}`

However, there is not much that Mathematica® can do with this kind of function. We cannot, for example, draw a graph of it since for plotting of graphs Mathematica® uses approximate numbers, and Mathematica® does not consider approximate numbers as either rational or irrational:

In[·]:= Element[1.1, Rationals]
Out[·]:= 1.1 ∈ Q

We can of course get some idea of the way the graph of such a function looks by using DiscretePlot. Here we combine two plots, one over a set of rationals and another one over irrationals.

In[·]:= Show[DiscretePlot[dirichlet[t], {t, -1, 1, 1/10},
          Filling -> None], DiscretePlot[dirichlet[t], {t, -1, 1,
          Sqrt[2]/15}, Filling -> None]]

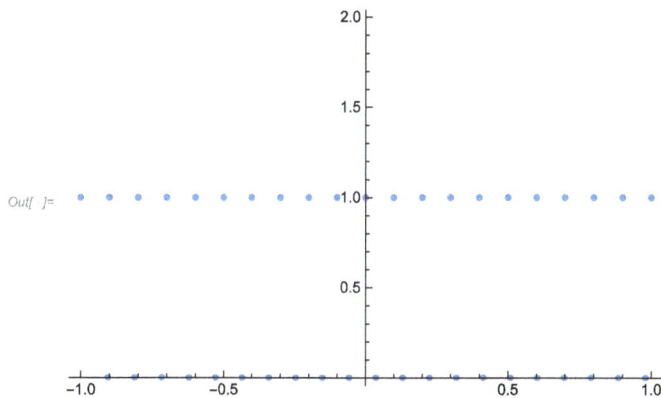

Figure 4.10

It is easy to see that the Dirichlet function is periodic and every rational number is a period.

In[·]:= $Assumptions = {Element[q, Rationals],
          Element[s, Rationals], NotElement[v, Rationals]}
Out[·]:= {q ∈ Q, s ∈ Q, u ∉ Q, v ∉ Q}

In[·]:= dirichlet[s + q]
Out[·]:= 1

In[·]:= dirichlet[v + q]
Out[·]:= 0

Similarly we can define a function that is continuous only at the point 0:

In[·]:= dd[x_] = Piecewise[{{x, Element[x, Rationals]}}]

$$Out[·]:= \begin{cases} x & x \in Q \\ 0 & \text{True} \end{cases}$$

In[·]:= Show[DiscretePlot[dd[t], {t, -1, 1, 1/10},
    Filling -> None], DiscretePlot[dd[t],
    {t, -1, 1, Sqrt[2]/15}, Filling -> None]]

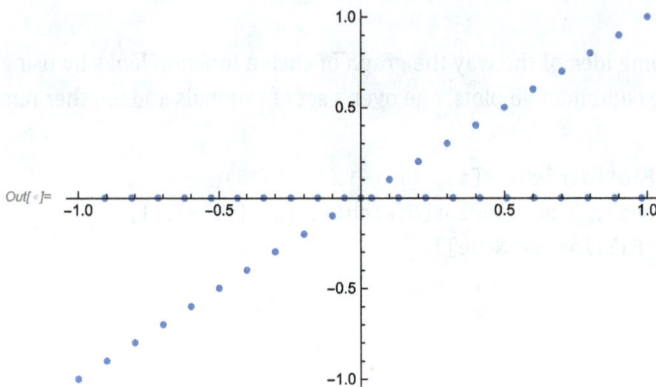

Figure 4.11

4.5 The main theorems on continuous functions

In this section we will discuss several important theorems on continuous functions.

Theorem 11 (The Weierstrass theorem). *Let I be a closed interval $[a, b]$. A continuous function $f : I \to \mathbb{R}$ has a maximum and a minimum value (i. e., it attains its supremum and infimum).*

Let I denote an interval (which could be infinite). We say that a function $f : I \to X$ has the Darboux property if for any a, $b \in I$ and any real number p between $f(a)$ and $f(b)$ there exists a real number q between a and b such that $f(q) = p$.

A discontinuous function can have the Darboux property; for example, the function

$$f(x) = \begin{cases} x, & x \le 1, \\ \sin(1/(x-1)), & x > 1, \end{cases}$$

which appeared in Section 4.4, clearly has the Darboux property although it is discontinuous at 1. However, we have the following theorem (see [14, Theorem 4.14]).

Theorem 12 (Intermediate value property). *A continuous function $f : I \to \mathbb{R}$ has the Darboux property.*

A form of the theorem (easily proved to be equivalent) that is very useful in applications is the following. If a and b are two unequal real numbers with $a < b$ and f is a continuous function on $[a, b]$ such that the values of f at a and b have opposite signs, then the equation $f(x) = 0$ has a solution in the interval $[a, b]$.

4.5.1 Example

Consider the following problem: find a solution of the equation $x^3 - 3x = -1$ with precision $1/100$ (i. e., a number x such that there is an exact root of the equation p with $|x - p| < 1/100$).

We consider the function

In[·]:= `Clear[f]; f[x_] := x^3 - 3*x + 1`

and look for solutions of $f(x) = 0$. Mathematica® has several built-in functions that can solve this problem with arbitrary precision (see below), but we shall use the Intermediate Value Theorem and a simple `While` loop. Before we start we need to find two real numbers a and b such that $f(a)$ and $f(b)$ have different signs. Without looking at the graph of f this needs some guessing, e. g.,

In[·]:= `f[0]`
Out[·]:= `1`

In[·]:= `f[1]`
Out[·]:= `-1`

So we know that there is a root between 0 and 1. We can now start at 0 and keep adding $1/100$ until we get to a number p such that $f(p)$ is negative. Then there has to be a root in the interval $(p - 1/100,\ p)$, which will be our solution.

In[·]:= `a = 0; While[f[a] > 0, a += 1/100]; {a - 1/100, a}`
Out[·]:= $\left\{\dfrac{17}{50},\ \dfrac{7}{20}\right\}$

In[·]:= `N[{17/50, 7/20}]`
Out[·]:= `{0.34, 0.35}`

We check the graph:

In[·]:= `Plot[f[x], {x, 17/50, 7/20}]`

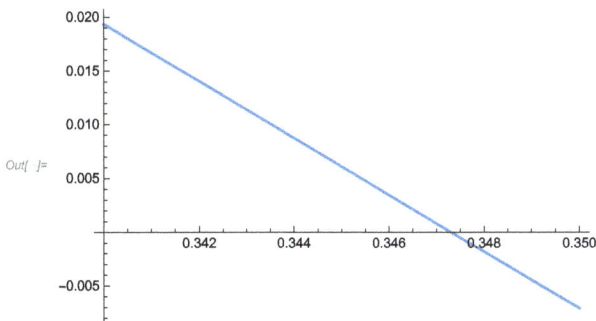

Figure 4.12

Thus any number in the interval (17/50, 7/20) is a solution to the problem.

There is another natural algorithm we can use. We start as above, by finding two points a and b (e.g., 0 and 1) where the function has different signs. We then divide the interval $[a, b]$ into two equal halves, $[a, c]$ and $[c, b]$. If the signs of f at a and c are different, we choose the interval $[a, c]$ otherwise we choose $[c, b]$. We then subdivide again and continue the process until the width of the interval (denoted below by d) is less than 1/100. We then return the midpoint:

> *In[·]:=* `Module[{a = 0, b = 1, c = 1/2, d = 1},`
> `    While[d >= 1/100, If[Sign[f[a]*f[c]] == -1,`
> `    a = a; b = c; c = (a + b)/2; d = b - a,`
> `    a = c; b = b; c = (a + b)/2; d = b - a]]; N[c, 3]]`
>
> *Out[·]:=* `0.348`

Of course Mathematica® has a number of built-in functions that can solve this problem (with the required precision or much higher if desired). As the equation is a polynomial one, we could use the function `NSolve`:

> *In[·]:=* `NSolve[f[x] == 0, x, WorkingPrecision -> 2.6]`
> *Out[·]:=* `{{x -> -1.88}, {x -> 0.347}, {x -> 1.53}}`

The function `FindRoot` works even with non-algebraic equations but needs a starting value:

> *In[·]:=* `FindRoot[f[x] == 0, {x, 0}, WorkingPrecision -> 2.6]`
> *Out[·]:=* `{x -> 0.347}`

Mathematica® can find the solutions of polynomial equations exactly and also with an arbitrary precision:

> *In[·]:=* `rts = x /. Solve[x^3 - 3*x == -1, x, Reals]`
> *Out[·]:=* `{Root[1 - 3*#1 + #1^3 & , 1], Root[1 - 3*#1 + #1^3 & , 2],`
> `    Root[1 - 3*#1 + #1^3 & , 3]}`

> *In[·]:=* `N[rts, 2.6]`
> *Out[·]:=* `{-1.88, 0.347, 1.53}`

Note that we can also find a rational approximation to the exact root above:

> *In[·]:=* `Rationalize[rts[[2]], 1/100]`
> *Out[·]:=* $\dfrac{5}{14}$

4.5.2 Example

Find $g(t) = \sup\{f(x), 0 \le x \le t\}$, where $f(x) = x^3 - 7x^2 - x + 20$.

The function MaxValue gives the maximum value of the function f with respect to x. Using this function one can define a new function which is $g(t) = \sup\{f(x), 0 \le x \le t\}$.

In[·]:= Clear[f, g]; f[x_] := x^3 - 7*x^2 - x + 20
In[·]:= g[t_] = MaxValue[{f[x], 0 <= x <= t}, x];

In[·]:= g[t]
Out[·]:= Piecewise[{{20, 0 <= t <= (1/2)*(7 + Sqrt[53])},
     {20 - t - 7*t^2 + t^3, t > (1/2)*(7 + Sqrt[53])}},
    -Infinity]

In[·]:= Plot[{f[x], g[x]}, {x, 0, 8}]

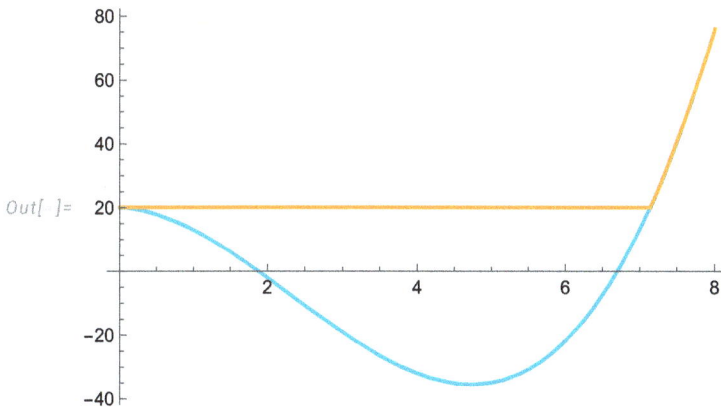

Out[ ]=

Figure 4.13

4.5.3 Example

The next example concerns functions defined as certain limits. The problem is to determine if these functions are continuous:

$$a) \quad f(x) = \lim_{n \to \infty} \frac{2n^x - n^{-x}}{n^x + n^{-x}}; \quad b) \quad g(x) = \lim_{n \to \infty} \frac{x^2 e^{(n+1)x}}{e^{nx} + 1}.$$

In the first case plotting the function indicates that it is not continuous:

In[·]:= f[x_] := Limit[(2*n^x - n^(-x))/(n^x + n^(-x)),
   n -> Infinity]

In[·]:= Plot[f[x], {x, -5, 5}]

Out[ ]=

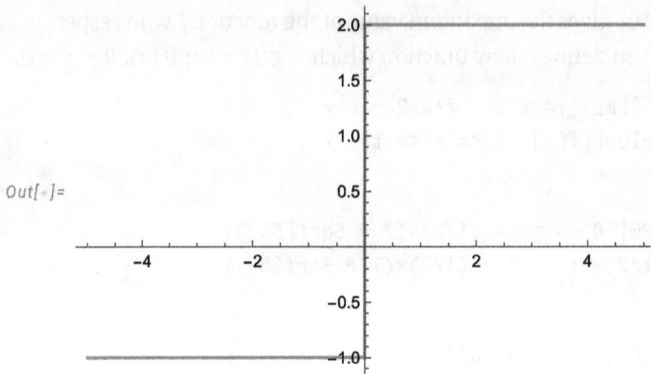

Figure 4.14

```
In[·]:= Limit[(2*n^x - n^(-x))/(n^x + n^(-x)),
        n -> Infinity]
Out[·]:= ConditionalExpression[2, x > 0]
```

Limit returns an incomplete answer. This does not, however, mean that the limit does not exist when the assumptions do not hold. To get the complete answer we need to divide the computation into three cases.

```
In[·]:= Limit[(2*n^x - n^(-x))/(n^x + n^(-x)),
        n -> Infinity, Assumptions -> x > 0]
Out[·]:= 2
```

```
In[·]:= Limit[(2*n^x - n^(-x))/(n^x + n^(-x)),
        n -> Infinity, Assumptions -> x < 0]
Out[·]:= -1
```

```
In[·]:= Limit[(2*n^x - n^(-x))/(n^x + n^(-x)),
        n -> Infinity, Assumptions -> x == 0]
Out[·]:= 1/2
```

Probably due to a bug, using ≥ instead of > gives a wrong answer:

```
In[·]:= Limit[(2*n^x - n^(-x))/(n^x + n^(-x)),
        n -> Infinity, Assumptions -> x >= 0]
Out[·]:= 2
```

Similarly, in the second case,

```
In[·]:= g[x_] := Limit[(x^2*E^((n + 1)*x))/(E^(n*x)
        + 1), n -> Infinity]
```

```
In[·]:= Plot[g[x], {x, -5, 5}]
```

Out[ ]=

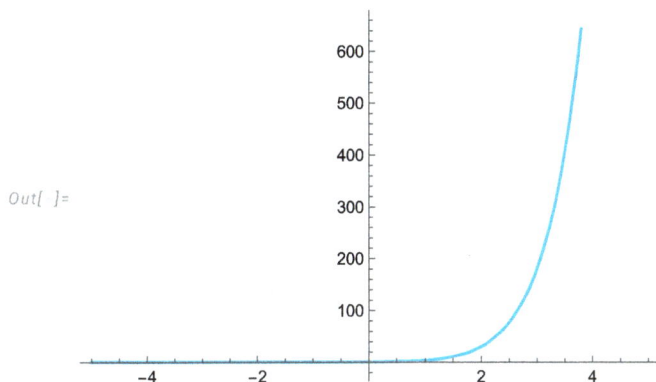

Figure 4.15

In[·]:= `Limit[(x^2*E^((n + 1)*x))/(E^(n*x) + 1),`
　　　　`n -> Infinity, Assumptions -> x < 0]`
Out[·]:= 0

In[·]:= `Limit[(x^2*E^((n + 1)*x))/(E^(n*x) + 1),`
　　　　`n -> Infinity, Assumptions -> x >= 0]`
Out[·]:= $E^x x^2$

4.6 Inverse functions and their continuity

We say that a function $f : A \rightarrow B$ is *injective* or *one-to-one* if $f(x) = f(y)$ implies $x = y$. We know that f is injective if and only if it has an inverse function $f^{-1} : f(A) \rightarrow A$ (where $f(A)$ is the image of A under f).

As mentioned above, it is possible to check injectivity, surjectivity and bijectivity of a univariate function over the reals in Mathematica®. For example,

In[·]:= `FunctionBijective[Sinh[x], x]`
Out[·]:= `True`

Let I be an interval. It follows easily from the intermediate value property theorem that a continuous function $f : I \rightarrow \mathbb{R}$ is injective if and only if it is either strictly monotone increasing ($x < y$ implies $f(x) < f(y)$) or strictly monotone decreasing ($x < y$ implies $f(x) > f(y)$). In this situation one can show that the inverse function is also continuous. This does not have to be true if the domain of f is not an interval. Consider, for example, the continuous functions $f : A \rightarrow B$ where $A = (0,1) \cup \{2\}$, $B = (0,1]$, $f(x) = x$ for $0 < x < 1$ and $f(2) = 1$. Then f has an inverse function $g = f^{-1}$, which is defined by $g(x) = x$, $x \in (0,1)$ and $g(1) = 2$, but it is not a continuous function (it does not have the Darboux property). We can illustrate this example as follows:

In[·]:= Graphics[{PointSize[0.02], Line[{{0, 0}, {1, 0}}],
    Point[{2, 0}], Line[{{0, 0}, {0, 1}}], Point[{0, 1}],
    Arrow[{{0.5, 0}, {0, 0.5}}], Arrow[{{0.98, 0},
    {0, 0.98}}], Arrow[{{2, 0}, {0, 1}}]}]

Out[·]=

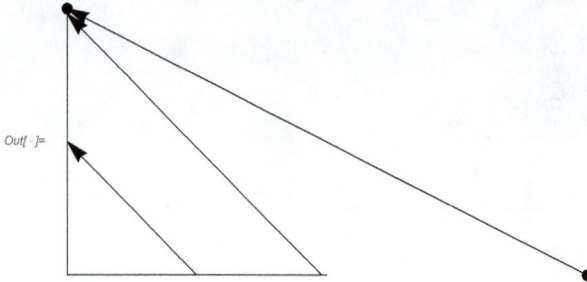

Figure 4.16

Mathematica® has a built-in function InverseFunction which will return the inverse of certain (not very complicated) functions:

In[·]:= InverseFunction[Sin]
Out[·]:= ArcSin

In[·]:= InverseFunction[Function[x, 3*x + 1]]
Out[·]:= Function[x, $\frac{1}{3}$ (-1 + x)]

Even when Mathematica® cannot give an explicit form of the inverse function, one can use it for computation and plotting:

In[·]:= f = Sin[#1] - #1 & ;
In[·]:= g = InverseFunction[f];
In[·]:= Plot[{f[x], g[x]}, {x, -2, 2}]

Out[·]=

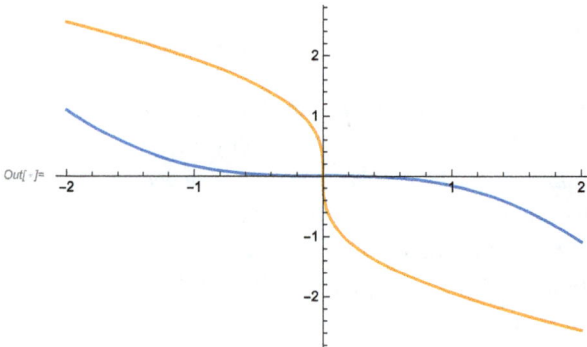

Figure 4.17

Mathematica® can compute symbolic inverse function:

In[·]:= InverseFunction[f][x]
Out[·]:= f^(−1)[x]

and automatically simplify the symbolic inverses:

In[·]:= f[InverseFunction[f][x]]

··· InverseFunction::ifun: Inverse functions are being used.

Values may be lost for multivalued inverses.

Out[·]:= x

When the function is not one-to-one, InverseFunction issues the same message as above, for example,

In[·]:= InverseFunction[#1^2 - 3*#1 + 5 &]

··· InverseFunction::ifun: Inverse functions are being used.

Values may be lost for multivalued inverses.

Out[·]:= (1/2)*(3 - Sqrt[-11 + 4*#1]) &

For functions with a named principal branch of the inverse, this warning message is not issued:

In[·]:= InverseFunction[Cos]
Out[·]:= ArcCos

The domain of the inverse function is computed automatically:

In[·]:= InverseFunction[ConditionalExpression[
        Sin[#1^2 - 1/2], 0 < #1 < 1] &]
Out[·]:= ConditionalExpression[Sqrt[1 + 2*ArcSin[#1]]/
        Sqrt[2], -Sin[1/2] < #1 < Sin[1/2]] &

As we already know, a closed-form representation for the inverse function might not exist:

In[·]:= f = InverseFunction[Sin[#1] - #1 &]
Out[·]:= InverseFunction[Sin[#1] - #1 &]

Evaluation of the inverse function at exact points yields exact numeric values:

In[·]:= f[2]
Out[·]:= Root[{-2 + Sin[#1] - #1 & , -2.55}]

However, the inverse may not be unique. In fact, Mathematica® returns three roots (the one above and two complex roots), which we omit:

In[·]:= Reduce[Sin[x] - x == 2 && Abs[x] < 5, x];

As discussed in "Functions That Do Not Have Unique Values" (tutorial/MathematicalFunctions#21968), many mathematical functions do not have unique inverses. In such cases, InverseFunction[f] can represent only one of the possible inverses for *f*.

4.7 Example: recursive sequences and continuity

We can now revisit the topic of recursive sequences. Suppose we are given a sequence defined by a system of recurrence equations,

$$a_1 = x, \quad a_{n+1} = f(a_n),$$

where f is a continuous function. Then if a_n is convergent to a limit $a \in \mathbb{R}$, then a has to be a fixed point of f. This follows immediately from the continuity of f since

$$f(a) = f\left(\lim_{n\to\infty} a_n\right) = \lim_{n\to\infty} f(a_n) = \lim_{n\to\infty} a_{n+1} = a.$$

Now let us assume that the function f has two fixed points a and b and it is defined and one-to-one (injective) on $[a, b]$. Then since f is monotonic, clearly $f([a, b]) = [a, b]$. In other words, if our recursive sequence starts at some x in between two fixed points, the whole sequence has to remain in between these two fixed points. Moreover, since an injective function has to be monotonic, one of these points has to be the limit, depending on whether the sequence is increasing or decreasing.

Let us illustrate this with an example. Consider the function given by

In[·]:= `Clear[f]; f[x_] := 3 - 1/x`

There are only two fixed points:

In[·]:= `x /. Solve[f[x] == x, x]`
Out[·]:= $\left\{\dfrac{1}{2}(3 - \sqrt{5}), \dfrac{1}{2}(3 + \sqrt{5})\right\}$

which means these are the only possible limits of the sequence

$$a_1 = x, \quad a_{n+1} = 3 - \frac{1}{a_n}.$$

Let us try to analyze the behavior of this sequence for various starting real numbers x. First we need to deal with the problem that for certain starting values x, such as $1/3$, a term of the sequence can become 0, which means that the subsequent terms will not be defined. We will say that the sequence explodes in such a case and we need to exclude such points. The set of points at which the sequence explodes satisfies an inverse recurrence relation:

In[·]:= `Solve[a[n + 1] == 3 - 1/a[n], a[n]]`
Out[·]:= $\left\{\left\{a[n] \rightarrow \dfrac{1}{3 - a[n + 1]}\right\}\right\}$

So we can define

$$b_1 = 0, \quad b_{n+1} = \frac{1}{3 - b_n}.$$

We can solve this with RSolve and find that we get an increasing sequence of points with limit

$In[\cdot]:=$ Limit[Simplify[b[n] /. RSolve[{b[n + 1] == 1/(3 - b[n]),
                b[1] == 0}, b[n], n]], n -> Infinity]

$Out[\cdot]:=$ $\left\{\frac{1}{2}\left(3 - \sqrt{5}\right)\right\}$

Let us now try to consider what will happen if we start at any other point. We know that the function must send the interval $[c_1, c_2] = [(3 - \sqrt{5})/2, (3 + \sqrt{5})/2]$ to itself. We only need to know where it is increasing:

$In[\cdot]:=$ Reduce[f[x] >= x, x]

$Out[\cdot]:=$ x < 0 $||$ $\frac{1}{2}(3 - \sqrt{5}) \leq x \leq \frac{1}{2}(3 + \sqrt{5})$

So the function is increasing in the interval $[c_1, c_2]$, hence if we start at any point of $(c_1, c_2]$, the limit must be c_2. Of course if we start at c_1 the sequence is constant. What about the other starting points? If we start to the right of c_2, then the function will be decreasing, but as it cannot "jump" into the interval or over it (because of continuity and injectivity), the sequence has to remain to the right of c_2 and converge to c_2 from the right. Moreover, if x is negative, $f(x)$ has to be positive and is greater than 3 (the "jump" is possible because f cannot be continuously defined at 0!), and after that the sequence will converge to c_2 from the right. Finally, if we start between 0 and c_1, then the sequence will decrease until it becomes negative (because it cannot be bounded below as otherwise it would have to have another limit), and then it will jump to the right side of c_2 and converge to it from above. The conclusion is that the sequence either explodes or converges to $c_2 = (3 + \sqrt{5})/2$, except when it starts at $c_1 = (3 - \sqrt{5})/2$ as it remains there. We suggest the reader to illustrate the behavior of this dynamical system by using Manipulate as in Chapter 2.

4.8 Uniform continuity and the Lipschitz property

A function $f : A \to \mathbb{R}$ is said to be *uniformly continuous* if for every $\varepsilon > 0$ there is a $\delta > 0$ such that for $x, y \in A$ with $|x - y| < \delta$ we have $|f(x) - f(y)| < \varepsilon$.

Note that when we write the definition in terms of quantifiers, the difference between ordinary continuity and uniform continuity amounts to the difference in the order of quantifiers: for continuity we have

$$\forall_{\varepsilon, \varepsilon > 0} \forall_{x, x \in A} \exists_{\delta, \delta > 0} \forall_{y, y \in A \wedge |x-y| < \delta} |f(x) - f(y)| < \varepsilon$$

and for uniform continuity we have

$$\forall_{\varepsilon, \varepsilon > 0} \exists_{\delta, \delta > 0} \forall_{x, x \in A} \forall_{y, y \in A \wedge |x-y| < \delta} |f(x) - f(y)| < \varepsilon.$$

Clearly uniform continuity implies ordinary continuity but the converse is not true. We can demonstrate this using Mathematica®. Since the condition is written entirely in terms of quantifiers, it should be possible to prove it for polynomial and rational functions by quantifier elimination. Let us compare two functions, the function

In[·]:= Clear[f, g]; f[x_] := x^2

defined on $\mathbb{R}$ and the function

In[·]:= g[x_] := Sqrt[x]

defined on $\mathbb{R}^+$. As we know both functions are continuous:

In[·]:= Reduce[ForAll[{e, x}, e > 0, Exists[d, d > 0,
          ForAll[{y}, Element[{y}, Reals] &&
          Abs[x - y] < d, Abs[f[x] - f[y]] < e]]]]
Out[·]:= True

In[·]:= Reduce[ForAll[{e, x}, e > 0 && x >= 0,
          Exists[d, d > 0, ForAll[{y}, y >= 0 &&
          Abs[x - y] < d, Abs[g[x] - g[y]] < e]]]]
Out[·]:= True

However, the first one is not uniformly continuous while the second one is:

In[·]:= Reduce[ForAll[e, e > 0, Exists[d, d > 0,
          ForAll[{x, y}, Element[{x, y}, Reals] &&
          Abs[x - y] < d, Abs[f[x] - f[y]] < e]]]]
Out[·]:= False

In[·]:= Reduce[ForAll[e, e > 0, Exists[d, d > 0,
          ForAll[{x, y}, x >= 0 && y >= 0 &&
          Abs[x - y] < d, Abs[g[x] - g[y]] < e]]]]
Out[·]:= True

We have the following theorem [14, Theorem 3.38].

Theorem 13. *Every continuous function f on a bounded closed interval [a, b] is uniformly continuous therein.*

We can demonstrate this by restricting the function $x \mapsto x^2$ to the unit interval:

In[·]:= Reduce[ForAll[e, e > 0, Exists[d, d > 0,
          ForAll[{x, y}, 0 <= x <= 1 && 0 <= y <= 1 &&
          Abs[x - y] < d, Abs[f[x] - f[y]] < e]]]]
Out[·]:= True

A function $f : A \to \mathbb{R}$ satisfies the *Lipschitz condition* if there exists a constant $c > 0$ such that $|f(x) - f(y)| \le c|x - y|$. Obviously, every function that satisfies the Lipschitz condition is uniformly continuous.

Let us show that the function $x \mapsto \sqrt{x}$ does not satisfy the Lipschitz condition on $[0, \infty)$ but does so on $[1, \infty)$:

In[·]:= Reduce[Exists[c, c > 0, ForAll[{x, y}, x >= 0 && y >= 0,
   Abs[Sqrt[x] - Sqrt[y]] <= c*Abs[x - y]]], Reals]

Out[·]:= False

In[·]:= Reduce[Exists[c, c > 0, ForAll[{x, y}, x >= 1 && y >= 1,
   Abs[Sqrt[x] - Sqrt[y]] <= c*Abs[x - y]]], Reals]

Out[·]:= True

In fact we can actually, instead of deciding whether such a constant c exists, find its value by removing the function Exists from the expression above:

In[·]:= Reduce[ForAll[{x, y}, x >= 1 && y >= 1,
   Abs[Sqrt[x] - Sqrt[y]] <= c*Abs[x - y]], Reals]

Out[·]:= $c \geq \dfrac{1}{2}$

Note that this time we needed to add the domain Reals to Reduce:

In[·]:= Reduce[Exists[c, c > 0, ForAll[{x, y}, x >= 0 &&
   y >= 0, Abs[Sqrt[x] - Sqrt[y]] <= c*Abs[x - y]]]]

 $\cdots$ Reduce: Reduce was unable to prove that a radical of
 an expression containing only real variables and parameters
 is real-valued. If you are interested only in solutions for
 which all radicals contained in the input are real-valued,
 use Reduce with domain argument Reals.

Out[·]:= $\text{Reduce}\left[\exists_{c,\,c>0}\forall_{\{x,y\},\,x\geq0\&\&y\geq0}\text{Abs}[\sqrt{x}-\sqrt{y}] \leq c\,\text{Abs}[x-y]\right]$

This happens whenever functions that can take non-real values, such as roots, appear in an expression to which Reduce is applied.

The reader should prove all the above results by hand. See also [14] for other examples.

4.8.1 Example

Finally, we consider an example which cannot be solved by using quantifier elimination (because it involves a non-algebraic function).

Consider the following problem: decide whether the function $\log(x)$ has the Lipschitz property or is uniformly continuous (a) on $(0, \infty)$ and (b) on $[a, \infty)$ for $a > 0$.

This is the kind of situation where Mathematica$^{\circledR}$'s dynamic graphic capabilities can be very useful. The dynamic graphic below shows that we can find $\varepsilon > 0$ (in this case $\varepsilon = 1$) and a pair of points $x, y > 0$ such that $|x - y| < \delta$ is arbitrarily small while $|\log(x) - \log(y)| = \varepsilon$. We can use the slider to make δ smaller and see that the distance between $\log(x)$ and $\log(y)$ (represented by the vertical green line) does not change (here $x = \delta$ and $y = \delta/e$). Hence, the function is not uniformly continuous on the interval $(0, \infty)$.

In[·]:= Manipulate[Show[Plot[Log[x], {x, 0.0001, 1.2},
    PlotRange -> {{0, 1.3}, {-5, 1}}],
    Graphics[{Red, Opacity[0.3], Rectangle[{d/E, 0},
    {d, Log[d/E]}], Opacity[1], Thickness[0.01],
    Blue, Line[{{d, 0}, {d, Log[d/E]}}], Line[{{d/E, 0},
    {d/E, Log[d/E]}}], Green, Line[{{d, Log[d]},
    {d, Log[d/E]}}]}], AxesOrigin -> {0, 0}, PlotRange ->
    {{0, 1.3}, {-5, 1}}], {{d, 1, "d"}, 0.001, 1}]

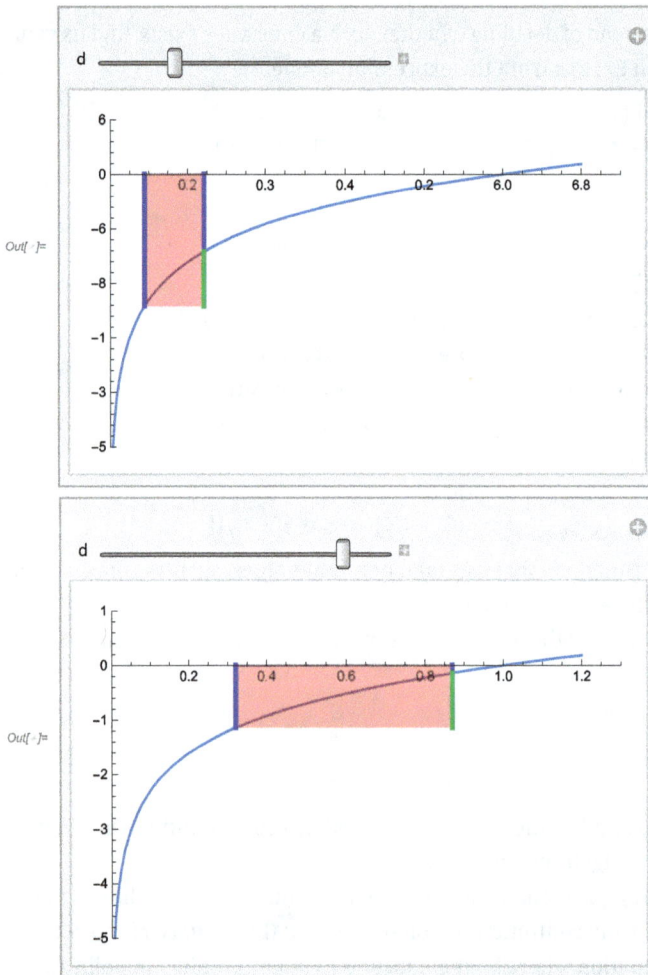

Figure 4.18

We can also see that if $x, y > a$ for some $a > 0$, then the above argument no longer works because we cannot move x and y arbitrarily close to 0. In fact, we can show that when restricted to $[a, \infty)$ the function log has the Lipschitz property. Indeed, suppose that $y > x$. Then

$$\log\left(\frac{y}{x}\right) = \log\left(\frac{x + (y - x)}{x}\right) = \log\left(1 + \frac{y - x}{x}\right) \leq \log\left(1 + \frac{y - x}{a}\right).$$

We now use the fact that $\log(z + 1) \leq z$ for $z \geq 1$, which Mathematica® can prove:

In[·]:= Reduce[Log[1 + z] <= z, z]
Out[·]:= z > -1

Hence

$$\log\left(\frac{y}{x}\right) \leq \frac{y - x}{a}$$

on $[a, \infty)$, which means that log has the Lipschitz property with the Lipschitz constant equal to $1/a$.

5 Differentiation

In this chapter we consider the concept of differentiability and derivative of a function at a point and on an interval. We see that differentiability is a special case of continuity, but of another function called the difference quotient. We then consider various properties of derivatives and applications of differentiation, in particular to finding global and local extrema and to convexity of functions. We also illustrate some aspects of the use of patterns and transformation rules in Mathematica®'s programming language by the example of a user-defined derivative.

5.1 Difference quotient and derivative of a function

Let I be an open interval in $\mathbb{R}$, $a \in I$ and let $f : I \to \mathbb{R}$ be a function. Its *difference quotient* is a function $Q : I \setminus \{a\} \to \mathbb{R}$ given by

In[·]:= Q[a_, f_][x] := DifferenceQuotient[f[x], {x, a - x}]

In[·]:= Q[a, f][x]

Out[·]:= $\dfrac{f[a] - f[x]}{a - x}$

Note that the difference quotient is not defined at the point a. We are going to consider the following question: can Q be extended to a function $\phi : I \to \mathbb{R}$ such that ϕ is continuous at a? If the answer is "yes", then we say that f is differentiable at a and $\phi(a)$ is its derivative at a.

In the interactive graphic below we show the graphs of the difference quotients of the functions $x \mapsto |x|^k$, where $k \in \{1/2, 1, 3/2, 2\}$. We see that the first two functions $\sqrt{|x|}$ and $|x|$ are not differentiable at 0 while the remaining functions are all differentiable.

In[·]:= Manipulate[Plot[Evaluate[Q[0, Abs[#1]^k &][x]],
        {x, -1, 1}, Exclusions -> 0, AxesOrigin -> {0, 0}],
        {{k, 1/2, "k"}, 1/2, 2, 1/2, Appearance -> "Labeled"},
        SaveDefinitions -> True]

https://doi.org/10.1515/9783111533063-005

Out[ ]=

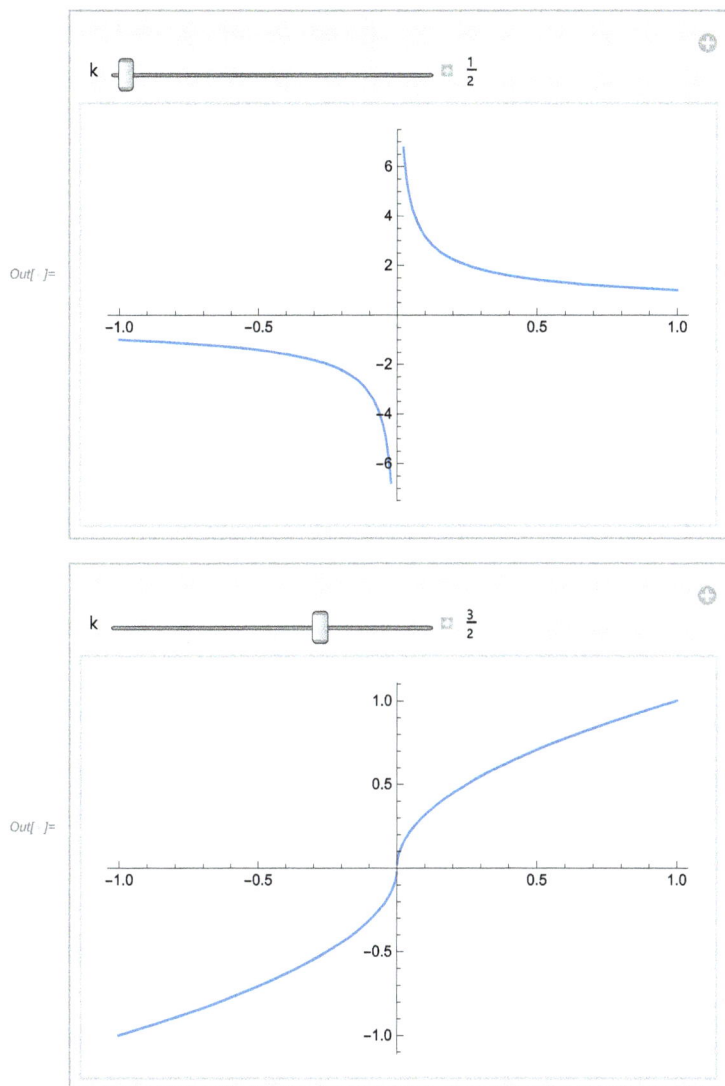

Out[ ]=

Figure 5.1

One can use the function DifferenceQuotient with the option Assumptions, for example:

In[·]:= DifferenceQuotient[Abs[x], {x, h}, Assumptions ->
       x < 0 && h < 0]

Out[·]:= -1

The definition can be expressed in the more usual way in terms of limits: f is *differentiable* at a if the limit

$$\lim_{x \to a} Q(a, f)(x) = \lim_{x \to a} \frac{f(x) - f(a)}{x - a}$$

exists, in which case the limit is called the *derivative* of f at a. The derivative of a function f at a point a is usually denoted as $f'(a)$ or $(\frac{df(x)}{dx})|_{x=a}$. If f is differentiable on an open interval I, then the derivative is also a function $f' : I \to \mathbb{R}, x \mapsto f'(x)$.

The second order derivative is defined at a point a whenever the first order derivative exists in a neighborhood of a and is differentiable (and hence also continuous) at a. We then say that f is twice differentiable at a and denote its second order derivative by $f''(a)$ or $(\frac{d^2 f(x)}{dx^2})|_{x=a}$. In general, we define the n-th order derivative of a function f as $f^{(n)} = (f^{(n-1)})'$. If f is n times differentiable and the n-th order derivative is continuous, then f is said to be of class C^n.

Geometrically the difference quotient represents the secant line of the graph of f connecting the points $(x, f(x))$ and $(a, f(a))$. As the point x approaches a, the secant approaches the tangent to the graph of f at $(a, f(a))$:

```
In[·]:= f[t_] := t^3
In[·]:= Manipulate[Show[Graphics[{Red, PointSize[0.01],
        Point[{1, 0}], Point[{1, f[1]}], Blue,
        Point[{1 + t, f[1 + t]}], Point[{1 + t, 0}],
        Thickness[0.005], Line[{{1, 0}, {1, f[1]}}],
        Line[{{1 + t, f[1 + t]}, {1 + t, 0}}],
        Line[{{1, If[t > 0, f[1], f[1 + t]]},
        {1 + t, If[t > 0, f[1], f[1 + t]]}}], Green,
        Line[{{1, f[1]}, {1 + t, f[1 + t]}}]],
        Plot[f[x], {x, 0, 2}], Plot[f[1] +
        Derivative[1][f][1]*(x - 1), {x, -1, 2},
        PlotStyle -> Red], Axes -> True,
        PlotRange -> {{0.3, 1.5}, {0, f[1.5]}},
        AspectRatio -> 1], {{t, 0.5, "h"}, -1, 1},
        SaveDefinitions -> True]
```

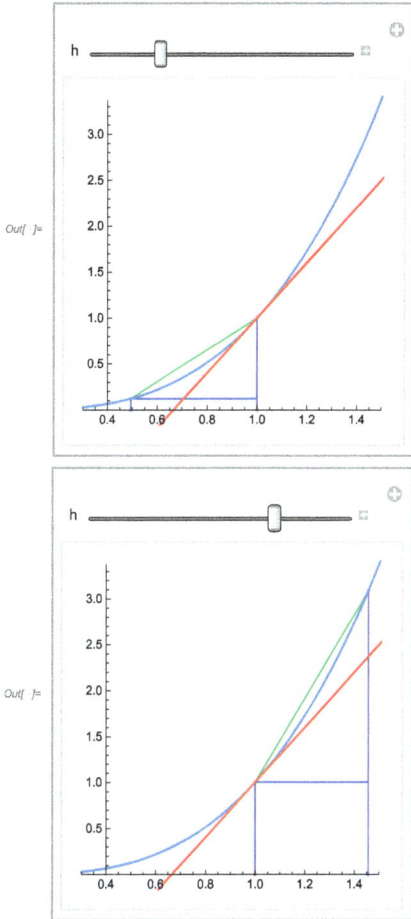

Figure 5.2

It is easy to see that if a function is differentiable at a point a, it has to be continuous there. Indeed, for any $a_n \to a$ we have

$$\lim_{n\to\infty} f(a_n) = \lim_{n\to\infty} \left(\frac{(a_n - a)(f(a_n) - f(a))}{a_n - a} + f(a) \right) = f(a).$$

When the limit $\lim_{x\to a^-} (f(x) - f(a))/(x - a)$ exists, the function f is said to be *left differentiable* at a and the limit is called its *left hand derivative*. If the limit $\lim_{x\to a^+} (f(x) - f(a))/(x - a)$ exists, then f is said to be *right differentiable* at a and the limit is its *right hand derivative*. A function is *differentiable* if and only if both its left and right hand derivatives exist and are equal.

The following is an example of a function which has both right and left hand derivatives at 1 but they are not equal (hence the function is not differentiable):

In[·]:= `Plot[Piecewise[{{x, x <= 1}, {x^2, x > 1}}], {x, 0, 2}]`

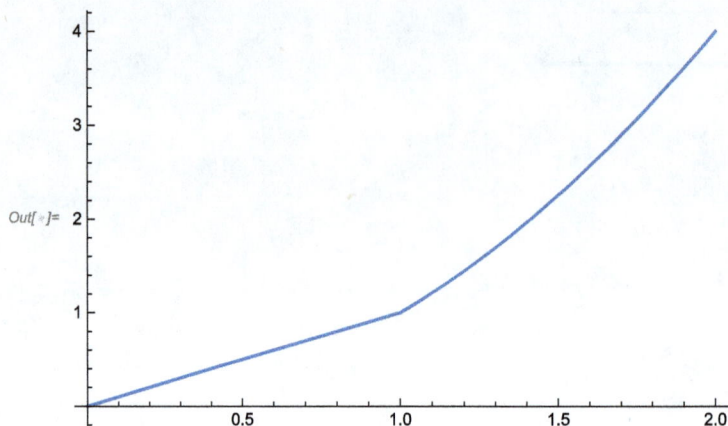

Figure 5.3

Indeed, the right hand derivative is

In[·]:= Limit[(Piecewise[{{x, x <= 1}, {x^2, x > 1}}]
        - 1)/(x - 1), x -> 1, Direction -> "FromAbove"]
Out[·]:= 2

while the left hand derivative is

In[·]:= Limit[(Piecewise[{{x, x <= 1}, {x^2, x > 1}}]
        - 1)/(x - 1), x -> 1, Direction -> "FromBelow"]
Out[·]:= 1

On the other hand, the function whose graph is shown below has neither left nor right hand derivatives:

In[·]:= Manipulate[Plot[Piecewise[{{x*Sin[1/x], x != 0}}],
        {x, -r, r}, PlotRange -> All], {{r, 1, "r"}, 0.01,
        1, Appearance -> "Labeled"}, SaveDefinitions -> True]

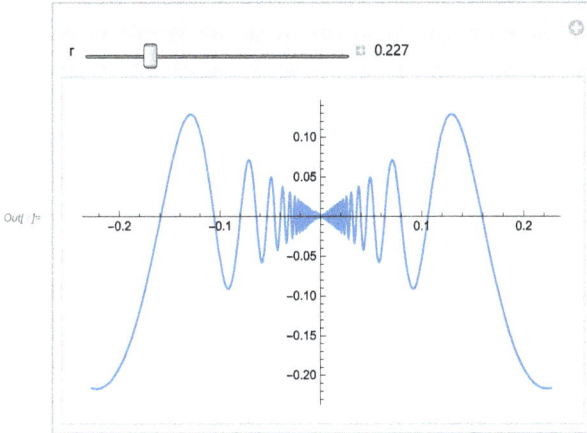

Figure 5.4

When the tangent to the graph at a point is vertical, the derivative at that point exists and is equal to ∞ but the function is said to be non-differentiable there:

In[·]:= `Plot[Surd[x, 3], {x, -1, 1}]`

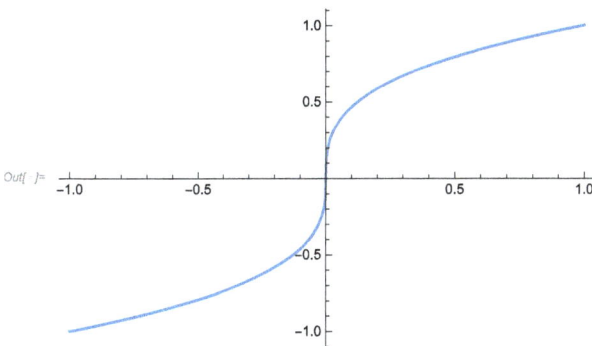

Figure 5.5

Note, however, that from the point of view of differential geometry the curve which represents the graph is regular (or smooth).

The reason why we use the function Surd rather than $x^{1/3}$ (Power[x, 1/3]) is that in Mathematica® $x^{1/n}$ is not a real number for negative x and odd integer n greater than one. For example,

In[·]:= `N[(-1)^(1/3)]`
Out[·]:= `0.5 + 0.866025 I`

This is due to the fact that in complex analysis any number has n complex roots. When n is odd, then any negative number always has just one real root, but this is not the "principal root", used in complex analysis. In the case $n = 3$ the principal root is a complex number with a positive imaginary part. In order to be consistent with this convention Mathematica®'s function Power always returns this (complex) principal root. Therefore, if one wants to get a real number, one needs to use the function Surd (or in this case the function CubeRoot), defined specifically for this purpose:

In[·]:= Surd[-1, 3]
Out[·]:= -1

In[·]:= CubeRoot[-1]
Out[·]:= -1

Note that in TraditionalForm the expression above will look like a cube root $\sqrt[3]{-1}$.

In the example below the derivative from the left is $-\infty$ and the one from the right is ∞. The tangent is also vertical but the derivative does not exist.

In[·]:= Plot[Sqrt[Abs[x - 2]], {x, 0, 4}]

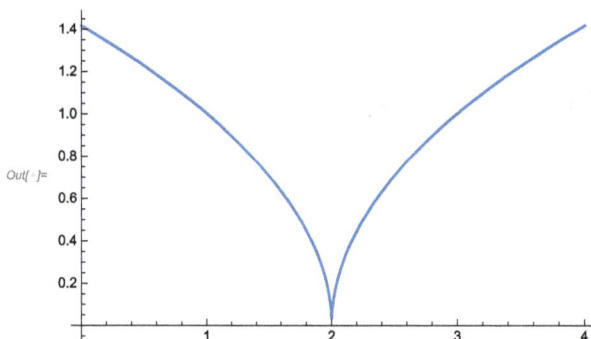

Out[·]=

Figure 5.6

Note that in this case the curve is not smooth.

Note that derivatives of order higher than the first are always defined inductively: that is, for $n > 1$, in order for a function to be n-times differentiable at a point, it must be differentiable $n - 1$ times in some neighborhood of this point. This is in contrast to what happens with the first derivative: for example, the function given by x^2dirichlet[x] is both continuous and differentiable only at the point $x = 0$. (Compute the limit of the difference quotient at 0.) However, the function x^3dirichlet[x] is not twice differentiable at $x = 0$, because for the second derivative to exist at a point, the first derivative must exist in an open neihgborhood of this point (and not just at the point).

The function dirichlet was defined in Section 4.4.1. Let us recall this definition:

In[·]:= dirichlet[x_] := Boole[Simplify[Element[x, Rationals]]]

In[·]:= Limit[DifferenceQuotient[x^2 dirichlet[x], {x, h}],
         h -> 0] /. x -> 0
Out[·]:= Limit[h*Boole[Element[h, Rationals]], h -> 0]

Although Mathematica® cannot see it by itself, this is, of course, 0.

In[·]:= Limit[DifferenceQuotient[x^3 dirichlet[x], {x, 2, h}],
         h -> 0] /. x -> 0 // Simplify
Out[·]:= Limit[6*h*Boole[Element[h, Rationals]], h -> 0]

This is also 0, but the funciton x^3 dirichlet[x] does not have the second order derivative at 0.

In general, if a function is n-times differentiable at x, its n-th derivative can be computed as a limit of a difference quotient. In Mathematica® the n-th difference quotient is computed as DifferenceQuotient[f[x],x,n,h]. For example:

In[·]:= DifferenceQuotient[f[x], {x, 2, h}]
Out[·]:= $\frac{f[x] - 2f[h+x] + f[2h+x]}{h^2}$

However, the limit of this expression will be equal to the second derivative only if the function f is twice differentiable in a neighbourhood of x. In Mathematica® we can use the option Analytic (which corresponds to the much stronger assumption that the function is analytic and, in particular, differentiable infinitely many times):

In[·]:= Limit[DifferenceQuotient[f[x], {x, 2, h}],
         h -> 0, Analytic -> True]
Out[·]:= f''[x]

We can easily construct an example of a function for which the limit of the n-th difference quotient exists but the function does not have the n-th derivative. For example, taking $n = 3$

In[·]:= g[x_] := Piecewise[{{x, x <= 0}, {x^2, x > 0}}]

In[·]:= Plot[g[x], {x, -2, 2}]

Out[ ]=

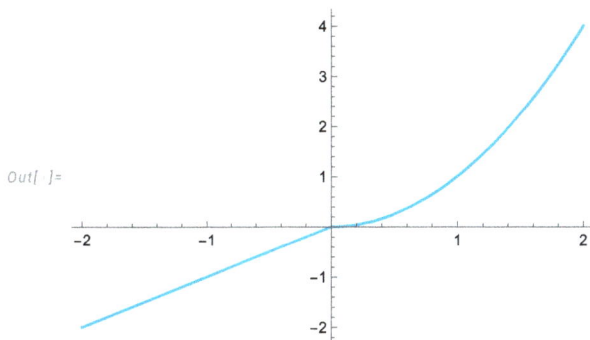

Figure 5.7

In[·]:= Limit[DifferenceQuotient[g[x], {x, 3, h}] /.
    x -> 0, h -> 0]
Out[·]:= 0

but

In[·]:= Derivative[3][g][0]
Out[·]:= Indeterminate

In this example the function does not have a second derivative at 0. It has both a left and right derivative, but their values do not agree:

In[·]:= Limit[DifferenceQuotient[g[x], {x, 2, h}] /.
    x -> 0, h -> 0, Direction -> "FromAbove"]
Out[·]:= 2

In[·]:= Limit[DifferenceQuotient[g[x], {x, 2, h}] /.
    x -> 0, h -> 0, Direction -> "FromBelow"]
Out[·]:= 0

5.2 Differentiation in Mathematica®

Mathematica®'s approach to differentiation is a little non-standard, since Mathematica® has two distinct notions of derivative, which we will call "derivative of an expression" and "derivative of a function". Each can be essentially used for the same purpose but they are conceptually different and each is more convenient in a certain context. It is therefore advisable to learn to use both, which is why we devote the whole section to this topic. In particular, we will explain why instead of writing

In[·]:= (x^2)'
Out[·]:= $(x^2)'$

we have to write either

In[·]:= D[x^2, x]
Out[·]:= 2 x

or

In[·]:= (Function[x, x^2])'
Out[·]:= Function[x, 2 x]

and instead of

In[·]:= Sin[x]'
Out[·]:= Sin[x]'

we must write

In[·]:= Sin'[x]
Out[·]:= Cos[x]

5.2.1 Differentiation of expressions using D

Recall that in Mathematica® "everything is an expression" and an expression always has the form $F[x_1, x_2, \ldots, x_n]$, where F is called the head of the expression and each x_i is an argument. Both the head and the arguments can themselves be either atoms or expressions of the above form. Here we will only consider certain special kinds of expressions which are related to mathematical functions. One can think of them as arising by applying a mathematical function to a certain number of variables. For example $x^2 + 2x + 1, 2x + a, \text{Sin}[x\,\text{Exp}[x\,y^2]] + 3$ are such expressions. It is important to keep in mind the distinction between expressions and the corresponding functions (of one variable); for example, the function Function$[x, x^2 + 2x + 1]$ corresponds to the expression $x^2 + 2x + 1$. In this section we will only consider functions of one variable, so if an expression contains two symbols, like $2x + a$, we think of a as a constant and take as the corresponding function Function$[x, 2x + a]$ (and not the function of two variables Function$[\{x, a\}, 2x + a]$).

Expressions are differentiated in Mathematica® using the symbol D. The derivative of an expression expr with respect to a variable x is written as D[expr,x]:

In[·]:= D[x*Sin[x], x]
Out[·]:= x Cos[x] + Sin[x]

Note that D treats every symbol in an expression that does not explicitly depend on x as a constant:

In[·]:= D[a*x^2 + b + 2, x]
Out[·]:= 2 a x

Otherwise if, for instance, we want b to depend on x, we need to write either

In[·]:= D[a*x^2 + b[x] + 2, x]
Out[·]:= 2 a x + b′[x]

or

In[·]:= D[a*x^2 + b + 2, x, NonConstants -> b]
Out[·]:= 2 a x + D[b, x, NonConstants -> {b}]

There is another approach (which we mention only briefly as we will not use it) that uses the "total differential" Dt instead of D. Dt makes the opposite assumption to that of D, i. e., that everything is non-constant unless declared to be a constant by means of the option Constants:

In[·]:= Dt[a*x, x]
Out[·]:= a + x Dt[a, x]

In[·]:= Dt[a*x, x, Constants -> a]
Out[·]:= a

In[·]:= Dt[a*x^2 + b, x, Constants -> {a, b}]
Out[·]:= 2 a x

If we want *a* to be treated as a constant globally, we can give it the Attribute Constant. The symbol *a* will then be treated by Dt as a constant until the kernel is quit or the Attribute is cleared:

In[·]:= SetAttributes[a, Constant]
In[·]:= Dt[a x, x]
Out[·]:= a
In[·]:= ClearAttributes[a, Constant]

If by using D we want to compute the value of the derivative of a function *f* at a point *a* we have to use a replacement rule, i. e., the following

In[·]:= Clear[f]; D[f[x], x] /. x -> a
Out[·]:= f'[a]

computes the derivative $f'(a)$. Note that the substitution of *a* for *x* has to be done after differentiation, otherwise we will get 0. Of course the name of the variable can be any allowed name, i. e., the expressions

In[·]:= D[x^3, x] /. x -> 2
Out[·]:= 12

In[·]:= D[var^3, var] /. var -> 2
Out[·]:= 12

both return the same answer. However, we have to be sure that the variable *x* does not have an assigned value or we can use a scoping construct such as Block or Module, e. g.,

In[·]:= x = 1; D[x^3, x]
       ··· General: 1 is not a valid variable.
Out[·]:= ∂_1 1

In[·]:= Block[{x}, D[x^3, x] /. x -> 2]
Out[·]:= 12
In[·]:= Clear[x]

In general when differentiating with respect to *x* an expression of the form F[x], where F is an atomic expression, Mathematica® treats the head of the expression as the name of a function and if it knows the derivative of F, for example G, then it returns G[x]:

In[·]:= D[Sin[x], x]
Out[·]:= Cos[x]

Otherwise it returns F'[x] (or in FullForm the output is Derivative[1][F][x]). The function Derivative will be discussed in Section 5.2.2.

In[·]:= D[Abs[x], x]
Out[·]:= Abs'[x]

In[·]:= % // FullForm
Out[·]//FullForm= Derivative[1][Abs][x]

Although Abs is a built-in function, Mathematica® does not have any value assigned as its derivative because in the complex plane Abs is not differentiable at any point, but in the real case we can give a formula which is valid for every point except 0. So we can define our own derivative that will suit our purpose:

In[·]:= Derivative[1][Abs] = Piecewise[{{1, #1 > 0},
          {-1, #1 < 0}}, Indeterminate] &

$$Out[·]:= \begin{cases} 1 & \#1 > 0 \\ -1 & \#1 < 0 \ \& \\ \text{Indeterminate} & \text{True} \end{cases}$$

Hence,

In[·]:= D[Abs[x], x]

$$Out[·]:= \begin{cases} 1 & x > 0 \\ -1 & x < 0 \\ \text{Indeterminate} & \text{True} \end{cases}$$

5.2.2 Differentiation of functions using Derivative

Derivative is used to differentiate functions. Mathematica® allows one to define functions in two ways: by means of patterns and rules and as pure functions. We can use Derivative in both cases.

We first define a function by means of patterns:

In[·]:= cube[x_] := x^3

Mathematica® remembers the following information about this function:

In[·]:= ?cube
Out[·]:= Global`cube
          cube[x_] := x^3

In[·]:= DownValues[cube]
Out[·]:= {HoldPattern[cube[x_]] :> x^3}

We can now differentiate the function cube by

In[·]:= Derivative[1][cube]
Out[·]:= 3 #1^2 &

The result is a pure function that can be immediately applied to arguments. There is a quicker way to achieve the same result:

In[·]:= cube′
Out[·]:= 3 #1² &

Note that we cannot use D in this case:

In[·]:= D[cube, x]
Out[·]:= 0

However, we can form an expression cube[x] which we can differentiate:

In[·]:= D[cube[x], x]
Out[·]:= 3 x²

Calculating derivatives at points using Derivative is very simple:

In[·]:= cube′[2]
Out[·]:= 12

Exactly the same approach works for pure functions, e. g.,

In[·]:= Function[x, x^3]′
Out[·]:= Function[x, 3 x²]

In[·]:= #^3 & ′[2]
Out[·]:= 12

The same method also works with built-in functions, for example:

In[·]:= {Sin′, Cos′}
Out[·]:= {Cos[#1] &, -Sin[#1] &}

In[·]:= Map[Derivative[1], {Sin, Cos}]
Out[·]:= {Cos[#1] &, -Sin[#1] &}

Note that the derivative of a constant function is zero:

In[·]:= Derivative[1][1 &]
Out[·]:= 0 &

Using Derivative has a number of advantages over using D; for example, it is much simpler to plot the graph of a derivative:

In[·]:= Plot[{cube[x], cube′[x]}, {x, -1, 1}]

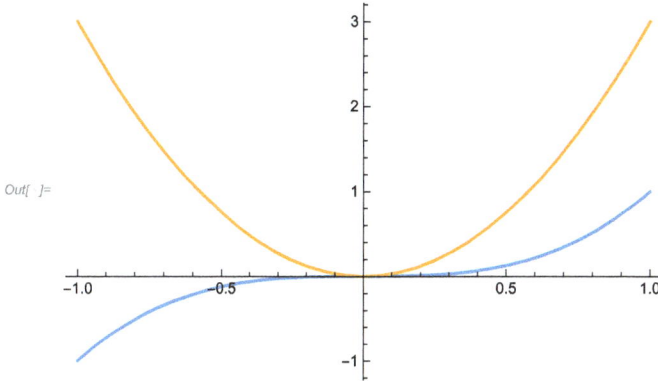

Out[ ]=

Figure 5.8

To achieve the same result by means of D we need to use a more complicated code:

In[·]:= Plot[Evaluate[{cube[x], D[cube[x], x]}], {x, -1, 1}]

In Mathematica® the *n*-th derivative of an expression f[x] is obtained by D[f[x], {x, n}] and of a function *f* as Derivative[n][f]. The last can also be entered for small *n* by using ' several times (for instance for $n = 2$ we write f'').

5.2.3 Algebraic rules of differentiation

The basic properties of derivatives follow from properties of limits and allow us to turn differentiation into an algorithm. More precisely, these properties show that once we know the derivatives of a certain family of "basic" functions there is an algebraic algorithm that computes the derivatives of their sums, products, inverses and compositions. Of course, the derivatives of these basic functions have to be computed using analytic means, that is, by computing suitable limits of difference quotient functions. The basic rules [14, Theorem 3.48] can easily be verified for the functions D or Derivative:

(i) The derivative of a constant is zero

In[·]:= D[1, x]
Out[·]:= 0

(ii) The sum rule

In[·]:= Clear[f,g]; D[f[x] + g[x], x]
Out[·]:= f'[x] + g'[x]

(iii) The Leibniz rule (the product rule)

In[·]:= D[f[x] g[x], x]
Out[·]:= g[x] f'[x] + f[x] g'[x]

(iv) The quotient rule

> *In[·]:=* D[f[x]/g[x], x] // Together
>
> *Out[·]:=* $\dfrac{g[x]\,f'[x] \,-\, f[x]\,g'[x]}{g[x]^2}$

The "chain rule" is the rule for computing the derivative of the composition of two functions [14, Theorem 3.51]:

> *In[·]:=* D[f[g[x]], x]
>
> *Out[·]:=* f′[g[x]] g′[x]

or

> *In[·]:=* Derivative[1][Composition[f, g]]
>
> *Out[·]:=* f′[g[#1]] g′[#1] &

From the chain rule we can derive the very important rule for differentiation of inverse functions:

> *In[·]:=* InverseFunction[f]′
>
> *Out[·]:=* $\dfrac{1}{f'\left[f^{(-1)}[\#1]\right]}$ &

or

> *In[·]:=* D[InverseFunction[f][x], x]
>
> *Out[·]:=* $\dfrac{1}{f'\left[f^{(-1)}[x]\right]}$

5.2.4 Example: user-defined derivative

A good exercise in Mathematica®'s pattern and rule-based programming is to write one's own version of D (or of Derivative):

```
In[·]:= d[Sin[x_], x_] := Cos[x]
        d[Cos[x_], x_] := -Sin[x]
        d[Exp[x_], x_] := Exp[x]
        d[Log[x_], x_] := 1/x
        d[f_, x_] /; FreeQ[f, x] := 0
        d[x_, x_] := 1
        d[(f_) + (g_), x_] := d[f, x] + d[g, x]
        d[(f_)*(g_), x_] := d[f, x]*g + f*d[g, x]
        d[(y_)^(a_), x_] := y^a*d[a*Log[y], x]
        d[(f_)[u_], x_] := Block[{v}, (d[f[v], v] /.
        v -> u)*d[u, x]]
```

The first four rules define how d operates on "basic functions". We have included only the functions Sin, Cos, Exp and Log. Next we have the rule that says that the derivative of a constant (i. e., an expression that does not explicitly involve x) is 0 and that the derivative of the identity function $x \mapsto x$ is 1. Next comes the rule for differentiating sums and products (the Leibniz rule). Next we have a rule for differentiating powers. In fact, our rule for differentiating powers is based on an easy to prove formula known as "logarithmic differentiation":

$$\frac{df(x)}{dx} = f(x)\frac{d\log(|f(x)|)}{dx}$$

(the formula is valid even without the absolute value if we accept the existence of logarithms of negative numbers, which we shall not go into here). The last rule is the chain rule.

Note that Mathematica® cannot "deduce" the rule for differentiating integer powers from the rule for differentiating products, like we do in ordinary mathematics, because Mathematica®'s rules apply to the FullForm of the expressions, and the FullForm of, for example, x^3 is

In[·]:= FullForm[x^3]
Out[·]//FullForm= Power[x, 3]

and not Times[x, x, x], which is the form of x x x before it gets evaluated:

In[·]:= FullForm[Hold[x x x]]
Out[·]//FullForm= Hold[Times[x, x, x]]

The reader can check that the derivative d we have just defined gives the same answers as the built-in function D, e. g.,

In[·]:= d[Sin[x] Log[x Cos[x^2]], x] // Simplify
Out[·]:= Cos[x] Log[x Cos[x²]] + Sin[x] $\left(\frac{1}{x} - 2x\,Tan[x²]\right)$

In[·]:= D[Sin[x] Log[x Cos[x^2]], x] // Simplify
Out[·]:= Cos[x] Log[x Cos[x²]] + Sin[x] $\left(\frac{1}{x} - 2x\,Tan[x²]\right)$

In[·]:= d[Sin[x]/x, x]
Out[·]:= $\frac{Cos[x]}{x} - \frac{Sin[x]}{x²}$

In[·]:= D[Sin[x]/x, x]
Out[·]:= $\frac{Cos[x]}{x} - \frac{Sin[x]}{x²}$

5.3 Main properties of differentiable functions

Let $f : I \to \mathbb{R}$, where I is an interval, and $a \in I$. We say that f has a *local maximum* (respectively *local minimum*) at a if there is some $\delta > 0$ such that for all $x \in I$ with

$|x - a| < \delta$ we have $f(x) \leq f(a)$ (respectively $f(x) \geq f(a)$). If strict inequalities hold we say that f has a strict local maximum (respectively strict local minimum) at a.

The word *extremum* refers to both maxima and minima. A global extremum is always a local one but, of course, the converse is not true. In fact, a differentiable function on a non-closed interval need not attain its global extrema (see the graph in Section 5.3.1).

The following simple fact is extremely useful.

Theorem 14 (Fermat's Interior Extremum Theorem). *Let I be an open interval, $a \in I$ and let $f : I \to \mathbb{R}$ be differentiable at a. If f has a local extremum at a, then $f'(a) = 0$.*

A point a such that $f'(a) = 0$ is called a *critical point* of f and $f(a)$ is called a *critical value*.

Suppose now that I is a closed interval and f is differentiable. Then by the Weierstrass theorem (see Theorem 11 in Chapter 4) the function f has to attain its supremum and infimum (i. e., it has to have a maximum and a minimum value). Thus the maxima must occur either at critical points or at the endpoints of the interval. In such cases we can find the global maximum and minimum by solving the equation $f'(x) = 0$, i. e., by finding the critical points and then comparing the critical values of f with its values at the endpoints. In general a differentiable function need not have a finite number of critical points. However, complex analytic functions (and in particular real ones) have a finite number of them, which follows from what is known as the Identity Theorem. If the number of critical points is finite, then we can find the global maximum and minimum by an algorithm (provided we have an algorithm for solving equations).

Another important theorem that we need is *Mean Value Theorem* [14, Theorem 4.17].

Theorem 15 (Mean Value Theorem). *If $f : [a, b] \to \mathbb{R}$, where $a < b$, is differentiable on (a, b) and is continuous on $[a, b]$, then there exists at least one point $c \in [a, b]$ such that*

$$\frac{f(b) - f(a)}{b - a} = f'(c).$$

This immediately implies that if $f'(c) > 0$ on an interval I, then f is strictly increasing on I and if $f'(c) < 0$, then f is strictly decreasing. The same holds with $>$ ($<$) replaced by $\geq$ ($\leq$) and the word "strictly" removed. However, if f is strictly increasing on I, we can only conclude that $f'(x) \geq 0$ on I (consider for example the function $f(x) = x^3$ on $[0, 1]$, which is strictly increasing, but $f'(0) = 0$).

Recall that a *Lipschitz function* $f : I = [a, b] \to \mathbb{R}$ was defined as one for which there exists a constant M such that $|f(x) - f(y)| \leq M|x - y|$ for all $x, y \in I$. A Lipschitz function is always continuous and even uniformly continuous but not necessarily differentiable. Suppose, however, that a function f is differentiable in (a, b). Then the Lipschitz condition is equivalent to the derivative of the function being bounded on I. Indeed, suppose that f is Lipschitz with constant M. Let $c \in (a, b)$. Then for every $x \in I \setminus \{c\}$

$$\frac{|f(x) - f(c)|}{|x - c|} \le M.$$

Taking limits we obtain $f'(c) \le M$, hence the derivative of f is bounded on I. Conversely, suppose that for all $x \in (a, b)$ we have $|f'(x)| \le C$. Let $x, y \in (a, b)$. Then by the Mean Value Theorem we have $(f(x) - f(y))/(x - y) = f'(\xi)$ for some $\xi \in (x, y) \subset (a, b)$. Hence $|f(x) - f(y)|/|x - y| = |f'(\xi)| \le C$ and f is Lipschitz with constant C. This makes it much easier to check if a differentiable function is Lipschitz. For example, consider the function $f(x) = \sin(\log(x))$ for $x \ge 1$. Since the function is differentiable, we need only to compute its derivative to show that it is Lipschitz:

In[·]:= D[Sin[Log[x]], x]

Out[·]:= $\dfrac{\text{Cos}[\text{Log}[x]]}{x}$

Clearly, on $[1, \infty]$ we have $|\cos(\log(x))|/|x| \le 1$, hence the function is Lipschitz.

Recall (see Section 4.6 in Chapter 4) that a continuous function $f : I \to \mathbb{R}$ defined on an interval has an inverse function (defined on its image) if and only if it is strictly monotone. The latter condition can be checked by computing the derivative, since we know that if $f'(x) > 0$ on I, then f is strictly monotone. If $f'(x) \ge 0$ on I, then f need not be strictly monotone (f can be constant on a non-trivial subinterval), and, hence, f may not have an inverse function. However, if in addition f is analytic and not constant, then $f'(x) \ge 0$ implies that f has an inverse function (for example, the inverse function to $f(x) = x^3$ is $g(x) = x^{1/3}$ on $\mathbb{R}$).

In[·]:= InverseFunction[#^3 &]

··· InverseFunction: Inverse functions are being used. Values

may be lost for multivalued inverses.

Out[·]:= #1$^{1/3}$ &

5.3.1 Example: global and local extrema

Next we consider a case that will demonstrate Mathematica®'s ability to solve problems that are very hard or impossible to solve by hand.

Let us find the maximum and minimum values of the function $f(x) = \sin(\cos(5x)) - e^{(x-1)^2}$ on the interval $[0, 2]$. We can plot the function to see the approximate answers:

In[·]:= f[x_] := Sin[Cos[5 x]] - E^(x - 1)^2

In[·]:= Plot[f[x], {x, 0, 2}]

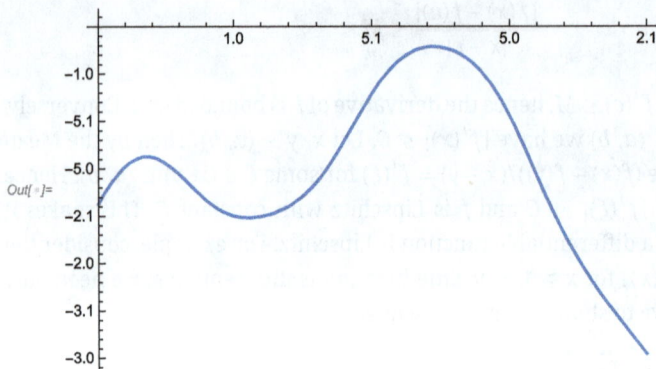

Figure 5.9

Mathematica®'s functions Maximize and Minimize can tell us the global maximum and minimum. Let us first compute the minimum:

In[·]:= Minimize[{f[x], 0 <= x <= 2}, x]
Out[·]:= {-E + Sin[Cos[10]], {x -> 2}}

The first element is the minimum value, the second one is a list {x− > a}, where a is a point where the minimum is attained. Note that Mathematica® returns only one such point, even though the function may attain minima at other points too. For example,

In[·]:= Minimize[(x - 1)^2 (x - 2)^2, x]
Out[·]:= {0, {x -> 1}}

but the function is also zero at point 2.

Returning to our example, we see that the minimum value is attained at the end of the interval, where $x = 2$ (as we can also see from the graph), and is equal to

In[·]:= f[2]
Out[·]:= -E + Sin[Cos[10]]

The answer given by Maximize will be more complicated, since Mathematica® needs to express the root in an exact form. There is no general standard form of expressing roots of such equations so Mathematica® does it using "root objects" (see Chapter 1):

In[·]:= Maximize[{Sin[Cos[5 x]] − E^(x − 1)^2, 0 <= x <= 2}, x]
Out[·]:= {−E^(−1+Root[{−2 E^(−1+#1)² +5 Cos[Cos[5 #1]] Sin[5 #1]+2 E^(−1+#1)² #1 &,1.22259830515136658892}])²

+ Sin[Cos[5 Root[
{−2 E^(−1+#1)² + 5 Cos[Cos[5 #1]] Sin[5 #1] + 2 E^(−1+#1)² #1 &,
1.22259830515136658892}]]],
{x → Root[{−2 E^(−1+#1)² + 5 Cos[Cos[5 #1]] Sin[5 #1] + 2 E^(−1+#1)² #1 &,
1.22259830515136658892}]}}

In[·]:= N[%]
Out[·]:= {-0.217221, {x -> 1.2226}}

The critical points of f can be found by solving the equation $f'(x) = 0$ in the interval $0 < x < 2$. This can be done either by using Solve or Reduce:

In[·]:= crits = x /. Solve[f'[x] == 0 && 0 < x < 2, x, Reals]
Out[·]:= {Root[{-2 E$^{-1+\#1^2}$ + 5 Cos[Cos[5 #1]] Sin[5 #1] + 2 E$^{-1+\#1^2}$
          #1 &, 0.180932451667257231810}],
          Root[{-2 E$^{-1+\#1^2}$ + 5 Cos[Cos[5 #1]] Sin[5 #1] + 2 E$^{-1+\#1^2}$
          #1 &, 0.55358202938476197032}],
          Root[{-2 E$^{-1+\#1^2}$ + 5 Cos[Cos[5 #1]] Sin[5 #1] + 2 E$^{-1+\#1^2}$
          #1 &, 1.22259830515136658892}]}

In[·]:= N[crits]
Out[·]:= {0.180932, 0.553582, 1.2226}

We see that there are three critical points. We can now compute the values of f at the critical points and at the endpoints. These points are our candidates for the maximum and the minimum, so we make a list of them:

In[·]:= candidates = (Join[{0}, crits, {2}]);
In[·]:= f /@ candidates // N
Out[·]:= {-1.87681, -1.37658, -2.02275, -0.217221, -3.4623}

We see that the maximum is attained at the fourth point and the minimum is at the last (the right endpoint 2).

Let us now consider the question of determining the nature of the critical points (that is, whether they are local maxima, local minima or neither). For this we will study the sign of the derivative of f in the interval (0, 2). Clearly, if a function is increasing (decreasing) to the left of a critical point and decreasing (increasing) to the right, the point is a local maximum (minimum). So we can determine the nature of the critical points by finding where the function is increasing and where it is decreasing. We can use Reduce to solve the inequality

In[·]:= Reduce[f'[x] > 0 && 0 < x < 2, x]
Out[·]:= 0 < x < Root{
          [-2 E$^{(-1+\#1)^2}$ + 5 Cos[Cos[5 #1]] Sin[5 #1] + 2 E$^{(-1+\#1)^2}$ #1 &,
          0.180932451667257231810}]||
          Root{[-2 E$^{(-1+\#1)^2}$ + 5 Cos[Cos[5 #1]] Sin[5 #1] + 2 E$^{(-1+\#1)^2}$ #1 &,
          0.55358202938476197032}] < x <
          Root{[-2 E$^{(-1+\#1)^2}$ + 5 Cos[Cos[5 #1]] Sin[5 #1] + 2 E$^{(-1+\#1)^2}$ #1 &,
          1.22259830515136658892}]

and we can immediately determine the nature of the critical points. However, there are also more elementary methods to demonstrate this. The first is to observe that the derivative

In[·]:= f′[x]

Out[·]:= $-2\,E^{(-1+x)^2}(-1 + x) - 5\,Cos[Cos[5\,x]]\,Sin[5\,x]$

is a continuous function. This implies (by the Darboux property) that the sign of the derivative can change only after passing through a critical point. This means that to determine the intervals of monotonicity we only need to check the signs of the derivative at points lying between the points (for example at midpoints). Below the function Partition divides our list of "candidates" into sublists of length 2 with offset 1. Applying the function Mean gives us a list of the midpoints and then by applying the function Sign we find the sign of the derivative at them:

In[·]:= Sign[f′[#]] & /@ (Mean /@ (Partition[candidates, 2, 1]))

Out[·]:= {1, -1, 1, -1}

From this we see that the first critical point is a local maximum, the second a local minimum and the third again a local maximum.

Note that we made the assumption that $f′$ is continuous. Is this always true? No, as we will show in Section 5.3.2. But it turns out that continuity of the derivative is not required since the derivative of any function always has the Darboux property, even when it is not continuous.[1]

The second approach to determine the nature of critical points makes use of the second order derivative. In the next chapter we will show (using Taylor's formula) that if the second order derivative at a critical point is positive, then the point is a local minimum and if it is negative, a local maximum. Let us check this in our case:

In[·]:= Sign[f″[#]] & /@ crits

Out[·]:= {-1, 1, -1}

We could try to compute the global maxima and minima on the whole $\mathbb{R}$. We first compute the limits as $x \to \infty(-\infty)$, which are both $-\infty$. This tells us that the function is not bounded from below. It also follows that there exists some interval $[-r, r]$ such that the maximum on this interval is equal to the maximum on $\mathbb{R}$ (in other words, the function is bounded above and its supremum is attained on $\mathbb{R}$). However, it is not in general possible to determine the value of r which is sufficient for this purpose. Moreover, as r increases finding the maximum value takes longer and longer.

5.3.2 Example

In the following example we will show that it is possible to have a differentiable function with a local minimum such that there is no interval to the left where the function is decreasing and no interval to the right where it is increasing. Moreover, we will show

---

1 https://en.wikipedia.org/wiki/Darboux%27s_theorem_(analysis)

that the derivative of this function is not continuous. Also we will be able to see that the Darboux property for the derivative holds.

We will define a function which depends on a parameter $c \geq 1$. We could define it as a function of two variables, x and c, but at this point we prefer to think of it as a function of one variable with a parameter.

In[·]:= `g[c_][x_] := Piecewise[{{(Sin[1/x] + c)*x^2,`
`        x != 0}, {c*x^2, x == 0}}]`

In[·]:= `Manipulate[Plot[g[c][x], {x, -r, r}, PlotRange ->`
`        All], {{c, 1, "c"}, 1, 2, Appearance -> "Labeled"},`
`        {{r, 1, "r"}, 0.01, 1, Appearance -> "Labeled"},`
`        SaveDefinitions -> True]`

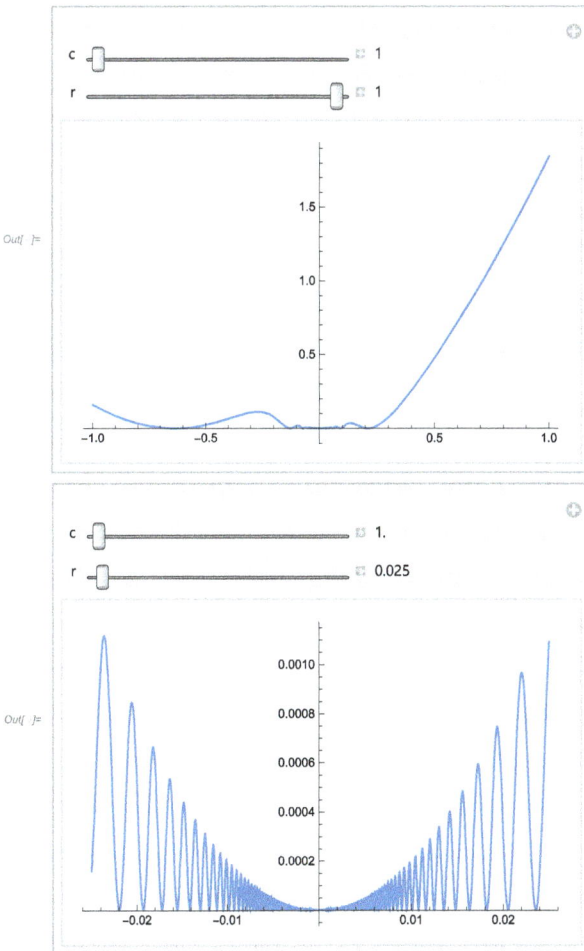

Figure 5.10

The function $g[c]$ has a local minimum at 0 (since $|\sin(1/x)| \leq 1$ and $c \geq 1$). For $c > 1$ the local minimum at 0 is strict. For $c = 1$ there are points arbitrarily close to 0 in which the function has the value 0.

Let us show that for every $r > 0$ the restrictions of $g[c]$ to $[0, r]$ and $[-r, 0]$ are not monotone. The derivative of $g[c]$ is given by

In[·]:= `FullSimplify[g[c]'[x]]`

Out[·]:= $\begin{cases} -\mathrm{Cos}\left[\dfrac{1}{x}\right] + 2x\left(c + \mathrm{Sin}\left[\dfrac{1}{x}\right]\right) & x \neq 0 \\ 0 & \text{True} \end{cases}$

Let us consider points $x_n = 1/(2\pi n)$, where n is an integer:

In[·]:= `Simplify[g[c]'[1/(2 Pi n)], Assumptions ->`
          `Element[n, Integers]]`

Out[·]:= $-1 + \dfrac{c}{n\pi}$

For large enough n these values are negative, while the points x_n are positive and as small as we like. Now consider the points $y_n = 1/(\pi(2n + 1))$, where n is an integer.

In[·]:= `Simplify[g[c]'[1/(Pi (2 n + 1))], Assumptions ->`
          `Element[n, Integers]]`

Out[·]:= $1 + \dfrac{2c}{(1 + 2n)\pi}$

This time by choosing large enough n we can make y_n arbitrarily close to 0 while $2c/(2\pi n + \pi) + 1 > 0$. The sequences $-x_n$ and $-y_n$ have analogous properties on the negative axis.

Let us now show that $g[c]$ is differentiable at 0 but its derivative is not continuous there. To see that the function $g[c]$ is differentiable, we consider the difference quotient defined earlier in this chapter:

In[·]:= `Q[0, g[c]][x]`

Out[·]:= $\begin{cases} x\left(c + \mathrm{Sin}\left[\frac{1}{x}\right]\right) & x \neq 0 \\ 0 & \text{True} \end{cases}$

In[·]:= `Limit[%, x -> 0]`
Out[·]:= `0`

Hence the derivative of $g[c]$ at 0 is defined and is 0. In order for the derivative to be continuous its limit as $x \to 0$ has to exist and has to be equal to the value of the derivative at 0. But the limit does not exist.

In[·]:= `Limit[-Cos[1/x] + 2*x*(c + Sin[1/x]), x -> 0]`
Out[·]:= `Interval[{-1, 1}]`

Mathematica® 11 somewhat surprisingly returns an interval rather than `Indeterminate`, although the limit of `Cos[1/x]` is `Indeterminate` and the limit of the second term is zero.

In any case the limit does not exist and the derivative is not continuous at zero. However, it is clear from the graph (for $c = 2$) that the derivative has the Darboux property:

In[·]:= `Plot[x*(2 + Sin[1/x]) - Cos[1/x], {x, -1, 1},`
         `Exclusions -> {x == 0}]`

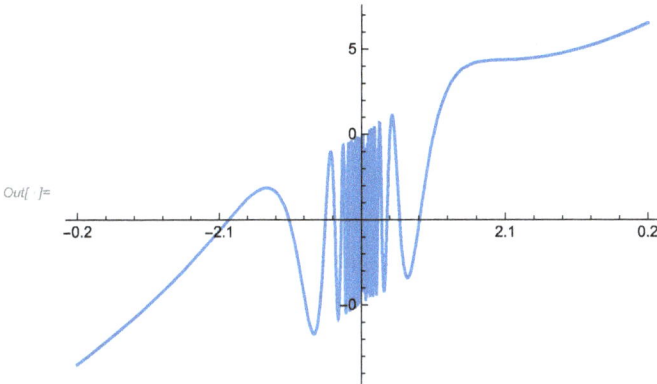

Out[·]=

Figure 5.11

5.3.3 Example: the inverse function of the hyperbolic sine

Let us consider the following example. We will consider the hyperbolic trigonometric functions, which in Mathematica® are defined as `Sinh` and `Cosh`. Since in this example we do not want to use Mathematica®'s built-in knowledge about these functions we will define our own versions:

In[·]:= `sinh[x_] := (Exp[x] - Exp[-x])/2`
In[·]:= `cosh[x_] := (Exp[x] + Exp[-x])/2`

Unfortunately in this example the function `InverseFunction` does not work correctly in Mathematica® 11.3 and 14.2:

In[·]:= `InverseFunction[sinh][x]`
        ··· InverseFunction: Inverse functions are being used. Values
        may be lost for multivalued inverses.
Out[·]= $\text{Log}\left[x - \sqrt{1 + x^2}\right]$

As we will see below the function has two inverses over the complex numbers and Mathematica® chooses the one which is incorrect over the real numbers. We can clearly see that this function does not take real values. It is a pity that the current version of Mathematica® does not allow to specify the domain over which `InverseFunction` should be considered. We will derive the correct answer below.

We will now show that the function sinh is invertible and compute the derivatives of its inverse. Our first approach is simply to find the inverse function. To show that the function sinh is invertible we need to show that the equation $\sinh(x) = y$ has at most one solution. In fact we will show that it has exactly one solution for every real y:

In[·]:= x /. Solve[y == sinh[x], x, Reals]
Out[·]:= $\{\text{Log}\left[y + \sqrt{1 + y^2}\right]\}$

This equation is easy to solve by hand (just use the substitution $e^x = t$ and solve the resulting quadratic equation). This means that the inverse function of sinh is $x \mapsto \log(x + \sqrt{x^2 + 1})$. We can now find its derivative directly:

In[·]:= Simplify[D[Log[Sqrt[x^2 + 1] + x], x]]
Out[·]:= $\dfrac{1}{\sqrt{1 + x^2}}$

The weakness of this approach is that it requires explicitly solving the equation $f(x) = y$ for x. In general this can be very hard to do or even impossible in an explicit form. However, we do not need to solve the equation to prove the existence of an inverse or even to find a formula for it. Since the function is defined on an interval, we only need to show that it is strictly monotone:

In[·]:= D[sinh[x], x]
Out[·]:= $\dfrac{1}{2}(E^{-x} + E^{x})$

Note also that the image of sinh is the whole of $\mathbb{R}$, since

In[·]:= FunctionRange[sinh[x], x, y]
Out[·]:= True

The answer True after applying the function FunctionRange means that every real number belongs to the range. Since the derivative is positive everywhere, the function sinh has an inverse function defined on $\mathbb{R}$. We can find the derivative using the formula

In[·]:= Derivative[1][InverseFunction[f]][x]
Out[·]:= $\dfrac{1}{f'\left[f^{(-1)}[x]\right]}$

Since the derivative of sinh is clearly cosh, this can be written as

$$\frac{1}{\cosh(\sinh^{(-1)}(x))}.$$

This is indeed a solution of the problem, but as it is not given in a convenient form, we now make use of the (easy to prove) identity $\cosh(x)^2 - \sinh(x)^2 = 1$ and obtain

$$\frac{1}{\cosh(\sinh^{(-1)}(x))} = \frac{1}{\sqrt{\sinh^2(\sinh^{-1}(x)) + 1}} = \frac{1}{\sqrt{1 + x^2}}.$$

Since cosh takes only positive values we take the positive square root. We can now check that this is exactly the answer we get by differentiating Mathematica®'s built-in function ArcSinh:

In[·]:= D[ArcSinh[x], x]

$$Out[·]:= \frac{1}{\sqrt{1 + x^2}}$$

5.3.4 Example: number of roots of an equation

Let us find the number of real roots of the equation $a^x = x$ for $a > 0$ and illustrate the answer using Mathematica®.

Mathematica® returns a complicated answer, involving unsolved equations (we omit the warning):

In[·]:= Clear[f, a]

In[·]:= FullSimplify[Reduce[a^x == x && a > 0, x,
        Reals], Assumptions -> a > 0]
Out[·]:= (E^ProductLog[-Log[a]]*x == 1 && (a < 1 ||
        1 < a <= E^(1/E)])) || (E^ProductLog[-1, -Log[a]]*x == 1 &&
        1 < a <= E^(1/E)]) || (a == 1 && x == 1)

Let us first look at some concrete numerical examples. This is always a good idea when working on a general problem with Mathematica®. We observe that there might be one or two roots or none at all for different values of the parameter a:

In[·]:= Solve[2^x == x, x, Reals]
Out[·]:= {}

In[·]:= Solve[(1/2)^x == x, x, Reals]
Out[·]:= {{x -> ProductLog[Log[2]]/Log[2]}}

In[·]:= Solve[(7/5)^x == x, x, Reals]
Out[·]:= {{x -> ProductLog[Log[5] - Log[7]]/(Log[5] - Log[7])},
        {x -> ProductLog[-1, Log[5] - Log[7]]/(Log[5] - Log[7])}}

We then study the problem graphically.

In[·]:= f[x_, a_] := a^x - x

In[·]:= Manipulate[Plot[f[x, a], {x, -r, r}, AxesOrigin
        -> {0, 0}, PlotRange -> All], {{a, 1, "a"}, 0, 4,
        Appearance -> "Labeled"}, {{r, 10, "range"}, 0.01,
        20, Appearance -> "Labeled"}, SaveDefinitions -> True]

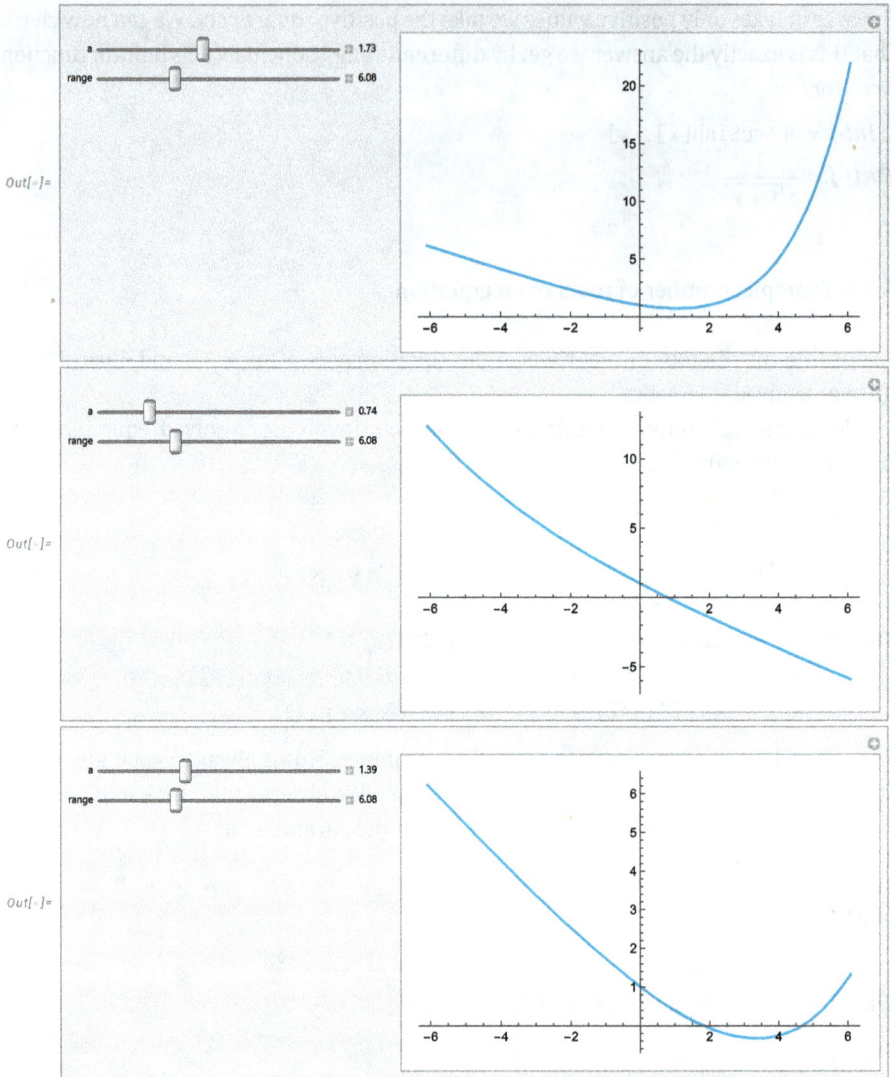

Figure 5.12

We see that for $a = 1$, there is only one solution, $x = 1$. Now let $0 < a < 1$. Let us examine the critical points of the function:

```
In[·]:= Solve[D[f[x, a], x] == 0 && 0 < a < 1, x, Reals]
Out[·]:= {}
```

We see that there are no critical points for such a. Moreover,

```
In[·]:= D[f[x, a], x]
Out[·]:= -1 + a^x Log[a]
```

For $a < 1$, since $\log(a) < 0$, the derivative is negative, and as

In[·]:= `Limit[f[x, a], x -> -Infinity, Assumptions -> 0 < a < 1]`
Out[·]:= `Infinity`

In[·]:= `Limit[f[x, a], x -> Infinity, Assumptions -> 0 < a < 1]`
Out[·]:= `-Infinity`

for $a < 1$ there is only one root.

For $a > 1$ we have

In[·]:= `Solve[D[f[x, a], x] == 0 && 1 < a, x, Reals]`
Out[·]:= `{{x -> ConditionalExpression[Log[1/Log[a]]/Log[a], a > 1]}}`

The value of the function at this point is

In[·]:= `FullSimplify[First[a^x - x /. %[[1]]], a > 1]`
Out[·]:= `(1 + Log[Log[a]])/Log[a]`

In[·]:= `D[f[x, a], {x, 2}]`
Out[·]:= `a^x*Log[a]^2`

The second order derivative is always positive, hence the derivative of the function is increasing. The function also has a local minimum at the point where the derivative vanishes. Hence, the equation has two roots if the value of the function at this minimum is negative.

In[·]:= `Reduce[1 + Log[Log[a]] < 0, a, Reals]`
Out[·]:= `1 < a < E^(1/E)`

In[·]:= `N[E^(1/E)]`
Out[·]:= `1.44467`

Hence, for $a = e^{1/e}$ there will be just one root, and for $a > e^{1/e}$ none.

5.3.5 Example: number of roots of an equation and the Lambert *W* function

We consider another, more complicated, example, which illustrates abilities of Mathematica® to make use of special functions to solve problems about elementary function. Consider the following problem of finding the number of roots of the equation

$$a^x = x^2$$

for various values of the parameter $a > 0$.

Mathematica® can solve the problem for a particular a:

In[·]:= `Solve[2^x == x^2, x, Reals]`
Out[·]:= `{{x -> 2}, {x -> -((2*ProductLog[Log[2]/2])/Log[2])},`
`        {x -> -((2*ProductLog[-1, -(Log[2]/2)])/Log[2])}}`

but cannot do this for an arbitrary a. Let us define a function g whose roots are the solutions of our equation:

In[·]:= g[x_, a_] := a^x - x^2

Mathematica® is unable to answer this directly:

In[·]:= Reduce[g[x, a] == 0, x, Reals]

··· Reduce::nsmet: This system cannot be solved with the methods available to Reduce.

Out[·]:= Reduce[a^x - x^2 == 0, x, Reals]

Over the complex numbers Mathematica® solves the equation in terms of the Lambert W function,[2] which is built-in, but it is not easy to determine which of the roots are real:

In[·]:= Reduce[g[x, a] == 0, x]
Out[·]:= (a == 1 && (x == -1 | x == 1)) || |
    (Element[C[1], Integers] && a != 0 && Log[a] != 0 &&
    (x == -((2*ProductLog[C[1], -(Log[a]/2)])/Log[a]) ||
    x == -((2*ProductLog[C[1], Log[a]/2])/Log[a])))

Here ProductLog[z] gives the principal solution for w in $z = we^w$. For example,

In[·]:= FullSimplify[ProductLog[z*Log[z]], Assumptions -> z > 1/E]
Out[·]:= Log[z]

since

In[·]:= z*Log[z] == w*E^w /. w -> Log[z]
Out[·]:= True

Mathematica® also cannot answer the question: for which values of a these solutions are real numbers?

In[·]:= FullSimplify[Element[2*ProductLog[C[1]],
    -(Log[a]/2)], Reals]]
Out[·]:= Element[ProductLog[C[1], -(Log[a]/2)], Reals]

We again can start by investigating the problem graphically, by using Manipulate:

In[·]:= Manipulate[Plot[g[x, a], {x, -r, r}, AxesOrigin
    -> {0, 0}, PlotRange -> All], {{a, 1, "a"}, 0, 4,
    Appearance -> "Labeled"}, {{r, 10, "range"}, 0.01,
    20, Appearance -> "Labeled"}, SaveDefinitions -> True]

---

2 https://en.wikipedia.org/wiki/Lambert_W_function

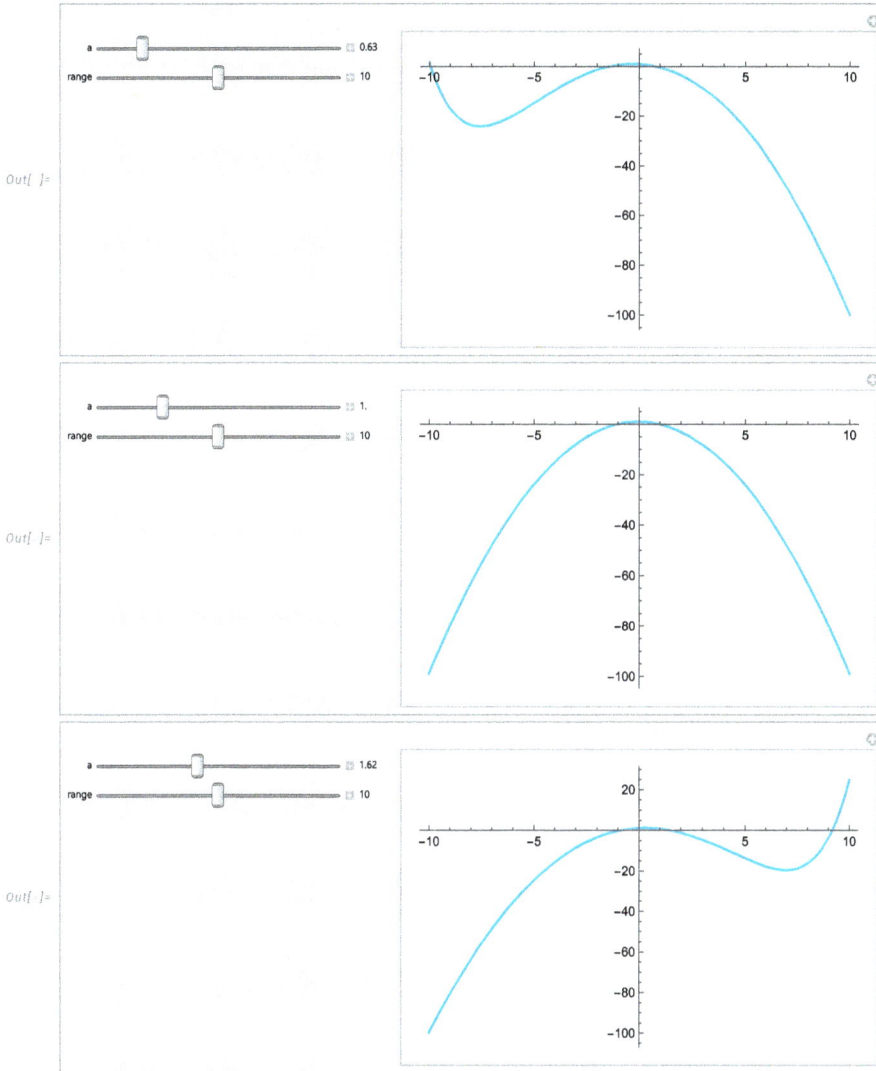

Figure 5.13

By varying a, we observe that, except the special case $a = 1$, when there are two solutions $x = 1$ and $x = -1$, the equation has either only one real root or three real roots. We also observe that in the situation when it has 3 roots, one of them lies between a local minimum and a local maximum. This suggests that we could try to solve the problem if we could find these local extrema. Fortunately, Mathematica® succeeds in doing this.

In[·]:= sols = Reduce[D[g[x, a], x] == 0 && a > 0, x, Reals]

$$\textit{Out[\cdot]=} \; (a = 1 \, \&\& \, x = 0) \, || \, \left(\text{Log}[a] \neq 0 \, \&\& \, \left(\left(e^{-\sqrt{\frac{2}{e}}} \leq a < 1 \, \&\& \, \left(x = -\frac{\text{ProductLog}\left[-\frac{1}{2}\,\text{Log}[a]^2\right]}{\text{Log}[a]} \, || \, x = -\frac{\text{ProductLog}\left[-1, -\frac{1}{2}\,\text{Log}[a]^2\right]}{\text{Log}[a]} \right) \right) \, || \right. \right.$$

$$\left. \left(1 < a \leq e^{\sqrt{\frac{2}{e}}} \, \&\& \, \left(x = -\frac{\text{ProductLog}\left[-\frac{1}{2}\,\text{Log}[a]^2\right]}{\text{Log}[a]} \, || \, x = -\frac{\text{ProductLog}\left[-1, -\frac{1}{2}\,\text{Log}[a]^2\right]}{\text{Log}[a]} \right) \right) \right) \right)$$

Figure 5.14

We observe that there is a special case, $a = 1$, when we know that the function is a quadratic with two roots and a global maximum at $x = 0$. In the cases $a < e^{-\sqrt{2/e}}$ and $a > e^{\sqrt{2/e}}$ there are no critical points, and by noting the limits of g as $x \to \infty$ and $x \to -\infty$, we see that there is only one real root.

In[·]:= `Limit[g[x, a], x -> Infinity]`
Out[·]:= `ConditionalExpression[Infinity, Log[a] > 0]`

In[·]:= `Limit[g[x, a], x -> -Infinity]`
Out[·]:= `ConditionalExpression[-\[Infinity], Log[a] > 0]`

To investigate the other cases we need to know the sign of the function g at the critical points. Let us find the values of g at these critical points and then use the Darboux property to show the existence of the root. The first value is given by

In[·]:= `crit1 = FullSimplify[a^x - x^2 /. x ->`
`        -(ProductLog[(-(1/2))*Log[a]^2]/Log[a]),`
`        Assumptions -> Inequality[E^(-Sqrt[2/E]),`
`        LessEqual, a, Less, 1]];`

and the second is

In[·]:= `crit2 = FullSimplify[a^x - x^2 /. x ->`
`        -(ProductLog[-1, (-(1/2))*Log[a]^2]/Log[a]),`
`        Assumptions -> Inequality[E^(-Sqrt[2/E]),`
`        LessEqual, a, Less, 1]];`

The answers are given in terms of the Lambert W special function, which we omit. We could try to learn more about this function but we can also see if Mathematica® can give us the answer itself. It turns out it can.

In[·]:= `Reduce[crit1 > 0, a]`
Out[·]:= $E^{-\sqrt{2/E}} \leq a < 1 || 1 < a \leq E^{\sqrt{2/E}}$

In[·]:= `Reduce[crit2 < 0, a]`
Out[·]:= $E^{-2/E} < a < 1 || 1 < a < E^{2/E}$

We can now see that for $E^{-\sqrt{2/E}} \leq a < e^{-2/e}$ and for $e^{2/e} < a \leq e^{\sqrt{2/e}}$ both critical values are positive and there is only one root. When $e^{-2/e} < a < 1$ or $1 < a < e^{2/e}$, one is negative and the other positive, and we have three roots.

5.3.6 Example: implicit differentiation

Find y' and y'' if the function $y = y(x)$ satisfies $x^2 + xy + y^2 = 3$.
 Let us define a function f by

In[·]:= `Clear[f]; f[x_, y_] := x^2 + x*y + y^2 - 3`

The zero set of this function is an ellipse.

In[·]:= `ContourPlot[f[x, y] == 0, {x, -3, 3}, {y, -3, 3}]`

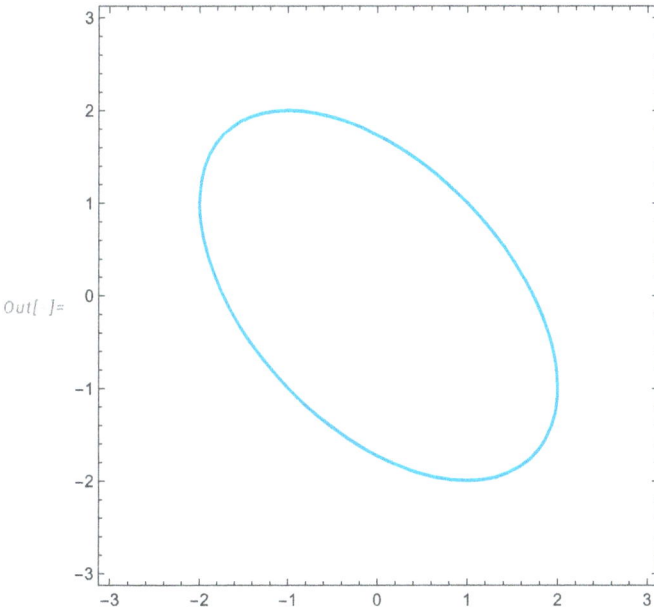

Out[ ]=

Figure 5.15

Let us find the first order derivative:

In[·]:= `Solve[D[f[x, y[x]] == 0, x], Derivative[1][y][x]]`
Out[·]:= `{{Derivative[1][y][x] -> (-2*x - y[x])/(x + 2*y[x])}}`

The second order derivative is

In[·]:= `Simplify[D[%, x] /. %[[1]]]`
Out[·]:= `{{Derivative[2][y][x] -> -((6*(x^2 + x*y[x] + y[x]^2))/`
`     (x + 2*y[x])^3)}}`

and after simplification

In[·]:= `Simplify[%, f[x, y[x]] == 0]`
Out[·]:= `{{Derivative[2][y][x] -> -(18/(x + 2*y[x])^3)}}`

Implicit differentiation can also be done using the built-in function ImplicitD:

In[·]:= ImplicitD[f[x, y] == 0, y, x]
Out[·]:= (-2*x - y)/(x + 2*y)

and

In[·]:= ImplicitD[f[x, y] == 0, y, {x, 2}]
Out[·]:= -((6*(x^2 + x*y + y^2))/(x + 2*y)^3)

5.3.7 Example: the Schwarzian derivative

Let us show that the function defined by

$$S(x) = \frac{x'''}{x'} - \frac{3}{2}\left(\frac{x''}{x'}\right)^2,$$

where $x = x(t)$, is invariant under the fractional-linear transformation

$$x \rightarrow \frac{ax + b}{cx + d},$$

where $ad - bc$ is not 0.

First we note that the condition on the coefficients a, b, c, d is necessary as otherwise

In[·]:= Simplify[(a*x[t] + b)/(c*x[t] + d) /.
        Solve[a*d - b*c == 0, d][[1]]]
Out[·]:= a/c

Define

In[·]:= S[x_, t_] := D[x, {t, 3}]/D[x, t] -
        (3/2)*(D[x, {t, 2}]/D[x, t])^2

In[·]:= FullSimplify[S[(a*x[t] + b)/(c*x[t] + d), t] -
        S[x[t], t]]
Out[·]:= 0

5.3.8 Example: a tangent line to a curve

Find the point of intersection of the tangent line to the curve $y = 3^x$ at the point $x = 1$ and the curve $y = x$.

Let

In[·]:= f[x_] := 3^x

The point of intersection of the tangent and the second curve is given by

In[·]:= a = x /. Solve[Derivative[1][f][1]*(x - 1) +
    f[1] == x, x][[1]]
Out[·]:= (3*(-1 + Log[3]))/(-1 + 3*Log[3])

In[·]:= N[%]
Out[·]:= 0.128858

In[·]:= Simplify[f[a]]
Out[·]:= 3^((3*(-1 + Log[3]))/(-1 + Log[27]))

In[·]:= Show[{Graphics[{Red, PointSize[0.04], Point[
    {{a, Derivative[1][f][1]*(a - 1) + f[1]}, {1,
    f[1]}}]}], Plot[{f[x], Derivative[1][f][1]*(x - 1) +
    f[1], x}, {x, -2, 2}]}]

Out[ ]=

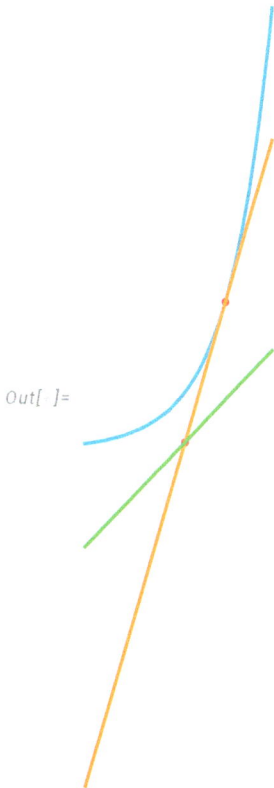

Figure 5.16

5.4 Convex functions

Definition 4. A function $f : I \to \mathbb{R}$ is said to be *convex* if for every $\lambda \in (0, 1)$ and every $x, y \in I$ we have

$$f(\lambda x + (1 - \lambda)y) \leq \lambda f(x) + (1 - \lambda)f(y).$$

If the above holds with $\leq$ replaced by $\geq$, then the function f is called *concave*.

Geometrically this corresponds to the fact, as illustrated below, that the blue point on the secant connecting two points with coordinates $(x, f(x))$ and $(y, f(y))$ lies above the red point on the graph of the function f:

```
In[·]:= Manipulate[Module[{f = (#1 - 1)^2 + 5 & },
        Show[Plot[f[x], {x, -1, 3}, AxesOrigin -> {0, 0}],
        Graphics[{Line[{{p[[1]], 0}, {p[[1]], f[p[[1]]]}}],
        Line[{{q[[1]], 0}, {q[[1]], f[q[[1]]]}}],
        Line[{{t*p[[1]] + (1 - t)*q[[1]], 0},
        {t*p[[1]] + (1 - t)*q[[1]], f[t*p[[1]] +
        (1 - t)*q[[1]]]}}],  Line[{{p[[1]], f[p[[1]]]},
        {q[[1]], f[q[[1]]]}}], Red, PointSize[0.02],
        Point[{t*p[[1]] + (1 - t)*q[[1]],
        f[t*p[[1]] + (1 - t)*q[[1]]]}], Blue,
        Point[{t*p[[1]] + (1 - t)*q[[1]], t*f[p[[1]]] +
        (1 - t)*f[q[[1]]]}]}]]], {{p, {0, 0}}, {-1, 0},
        {1.9, 0}, Locator}, {{q, {2, 0}}, {2, 0}, {3, 0},
        Locator}, {{t, 0.5, "t"}, 0, 1}, SaveDefinitions -> True]
```

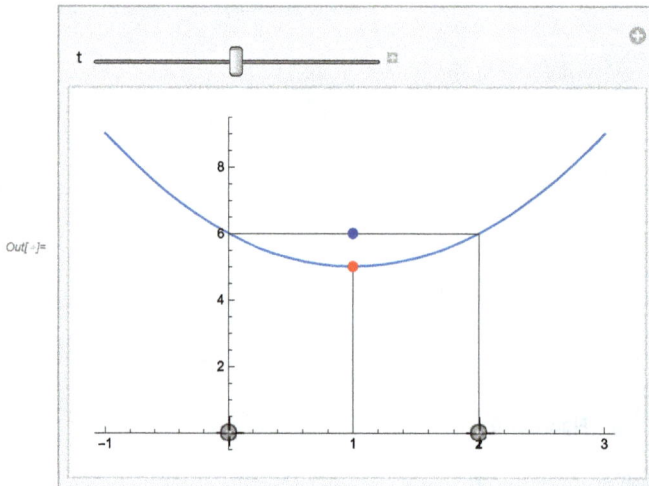

Figure 5.17

There is an equivalent formulation of convexity, which also has a geometric inter-pretation. Consider the following difference quotient function (of two variables):

$$\tilde{Q}(f)(x,y) = \frac{f(x) - f(y)}{x - y},$$

where $x, y \in I$. If we fix one of the variables, say y, we obtain a function of one variable (whose domain is $I \setminus \{y\}$). One can show that f is convex if and only if the difference quotient function is increasing when viewed as a function of one variable.

The following interactive illustration shows that the slope of any secant of the graph of a convex function is an increasing function:

```
In[·]:= Manipulate[Module[{f = (#1 - 1)^2 + 5 & ,
         g, P, Q}, g = Plot[f[x], {x, -1, 5}];
         P = {p[[1]], f[p[[1]]]}; Q = {q[[1]], f[q[[1]]]};
         Show[g, Graphics[{Text["P", P], Text["Q", Q],
         Red, Line[{P, Q}]}], Axes -> True, Frame ->
         False, PlotRange -> All, AxesOrigin -> {0, 0},
         Ticks -> None]], {{p, {0, 0}}, {-1, 0}, {1.9, 0},
         Locator}, {{q, {2, 0}}, {2, 0}, {5, 0}, Locator},
         SaveDefinitions -> True]
```

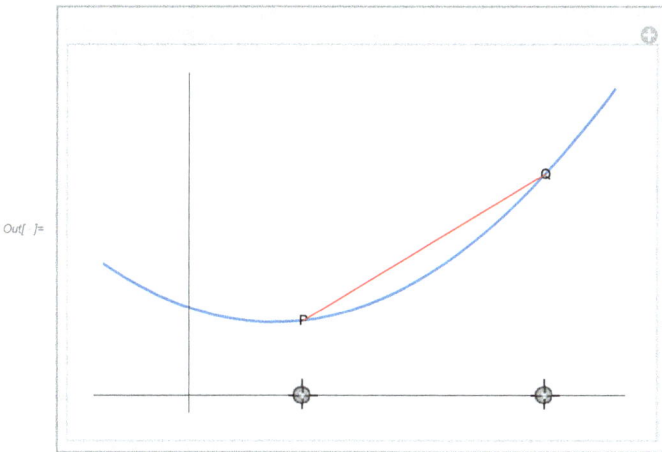

Figure 5.18

A convex function need not be differentiable or even continuous. For example it is easy to check that the following function is convex on $[-1, 1]$:

```
In[·]:= gg[x_] := Piecewise[{{2, x == -1},
         {Abs[x], -1 < x < 0}, {x^2, Inequality[0, LessEqual,
         x, Less, 1]}, {2, x == 1}}]
```

In[·]:= Plot[gg[x], {x, -1, 1}, Prolog ->
    {Point[{{-1, 3/2}, {1, 3/2}}]}, PlotRange ->
    {{-1, 1.1}, {0, 2}}]

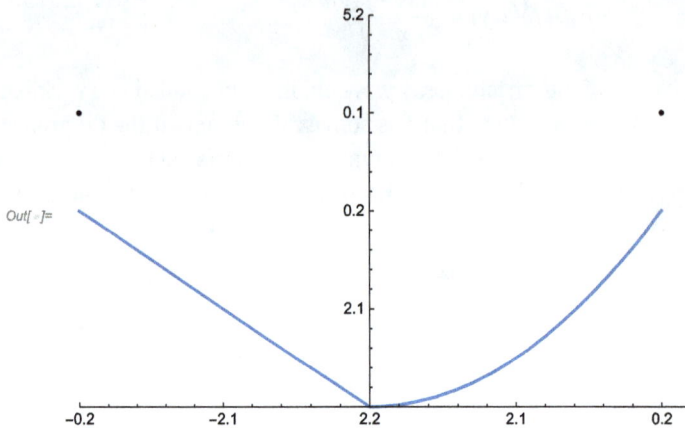

Out[·]=

Figure 5.19

This function is not differentiable at 0 (but is both left and right differentiable) and it is not continuous at the endpoints of the interval.

Using the fact that the difference quotient function of a convex function is increasing one can show that a convex function on an interval has both left and right derivatives at every interior point of the interval (because as we move the left point of the secant keeping the right point fixed, the slope of the secant gives us an increasing and bounded function, hence, the left derivative exists; the analogous argument holds for the right derivative). These two one-sided derivatives need not be equal but their existence implies that the function is continuous at all points in the interior of the interval (left [right] differentiability implies left [right] continuity which together imply continuity).

If f is differentiable (in the interior of its domain), then one can characterize its convexity in terms of the first order derivative. A geometric characterization is that the tangent to the graph of a convex function (at points over the interior of the domain) lies below the graph. Naturally, the tangents to the graph of a concave function lie above it. The function shown below is convex on certain subintervals of its domain (the convex part of the function is shown in red) and concave at others. The points where the behavior changes are endpoints of the intervals of convexity and concavity and the tangents at such point lie above the graph on one side and below on the other.

In[·]:= `Manipulate[Module[{f = Sin[6*#1^3] & , u, v},`
`        u = Minimize[{f[x], -1 <= x <= 1}, x][[1]];`
`        v = Maximize[{f[x], -1 <= x <= 1}, x][[1]];`
`        Show[Plot[f[x], {x, -1, 1}, ColorFunction ->`
`        Function[x, If[Derivative[2][f][x] > 0, Red, Blue]],`
`        ColorFunctionScaling -> False, Epilog ->`
`        {PointSize[0.02], Point[{a, 0}]}, PlotRange ->`
`        {{-1, 1}, (4/3)*{u, v}}], Plot[f[a] +`
`        Derivative[1][f][a]*(x - a), {x, -1, 1},`
`        PlotRange -> {{-1, 1}, (4/3)*{u, v}}]]],`
`        {{a, 0, "a"}, -1, 1, Appearance -> "Labeled"},`
`        SaveDefinitions -> True]`

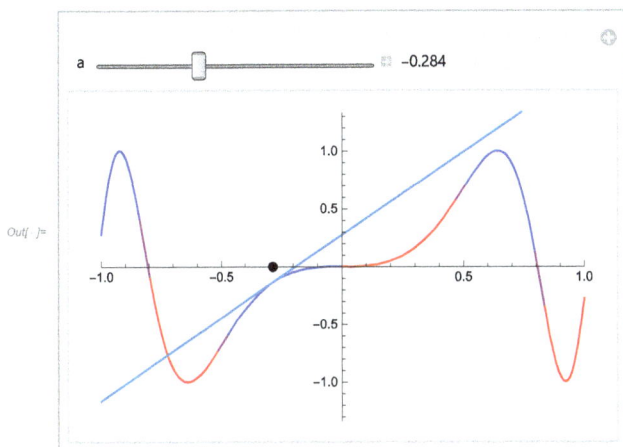

Figure 5.20

Another characterization of convexity of a differentiable function is that its derivative is increasing (the slopes of tangents are increasing). The derivative of a function is not necessarily differentiable itself, but when it is, the condition that the derivative is increasing can be replaced by the condition that the second order derivative is non-negative (positive for strict convexity). This is generally the most convenient condition and was used in the graph above to color the convex and concave parts of the graph red and blue (by means of the ColorFunction option of Plot). The option ColorFunctionScaling is used to stop Mathematica® from scaling the argument in ColorFunction to lie between 0 and 1.

From version 12.2 Mathematica® has built-in functions that can check the monotonicity and convexity of a given function (see guide/FunctionProperties). FunctionMonotonicity returns 1 for a function that is non-decreasing over the reals or in a specified region, and −1 if the function is non-increasing. For a constant function, 0 is returned. In other cases Indeterminate is returned.

In[·]:= FunctionMonotonicity[E^x, x]
Out[·]:= 1

FunctionConvexity returns 1 for a function that is convex over reals or in a specified region, and –1 if the function is concave. For an affine function, 0 is returned. In other cases Indeterminate is returned.

In[·]:= FunctionConvexity[Abs[x], x]
Out[·]:= 1

In[·]:= FunctionConvexity[Sin[x], x]
Out[·]:= Indeterminate

The following functions are concave for positive *x*:

In[·]:= FunctionConvexity[{Sin[x], 0 <= x <= Pi/2}, x]
Out[·]:= -1

In[·]:= FunctionConvexity[{Log[x], x > 0}, x]
Out[·]:= -1

FunctionConvexity has several useful options

In[·]:= Options[FunctionConvexity]
Out[·]:= {Assumptions :> $Assumptions, GenerateConditions
-> True, PerformanceGoal :> $PerformanceGoal,
StrictInequalities -> False}

To check strict convexity of a function one can use the option StrictInequalities:

In[·]:= FunctionConvexity[x^2*Abs[x], x,
StrictInequalities -> True]
Out[·]:= 1

FunctionConvexity can give a conditional answer:

In[·]:= FunctionConvexity[{x^4 + a*x^2, x > 1}, x]
Out[·]:= ConditionalExpression[1, Element[a,
Reals] && a >= 0]

Note that FunctionConvexity returns only sufficient conditions:

In[·]:= FunctionConvexity[{x^4 - x^2, x > 1}, x]
Out[·]:= 1

One can also check convexity for particular values of a parameter:

In[·]:= FunctionConvexity[x^4 + a*x^2, x, Assumptions
-> a < 0]
Out[·]:= Indeterminate

A classical way to define a convex function over an open interval is in terms of its lines of support.[3] Let $f : I \rightarrow \mathbb{R}$ be a function, where I is an open interval and let $x_0 \in I$. The line given by the equation $y(x) = f(x_0) + m(x - x_0)$ is called a line of support of f and x_0 if $y(X) \leq f(x)$ for all $x \in I$. A function is convex if it has at least one line of support at each $x \in I$. The set of all such linear functions at x_0 is called the subdifferential[4] of f at x_0. One can show that the value of a convex function at a point x is the supremum of all the values of the functions belonging to the subdifferentials of f at all points of I. One can also define convex functions on open intervals by starting with a family of lines with the property that supremum of their values at each point is finite. For finite families of lines we can implement this in Mathematica®. Below we construct a family of lines with random gradients and y-intercepts. The function that assigns to each point of the domain the y-coordinate of the highest point above it lying on one of the lines is convex.

```
In[·]:= convex[n_] := Module[{x, ll, f}, ll =
        RandomReal[{-20, 5}, n]*x + RandomReal[
        {-10, 80}, n]; f[a_] := Max[ll /. x -> a];
        Show[Plot[ll, {x, -10, 10}], Plot[f[a],
        {a, -10, 10}, PlotStyle -> {Thickness[0.01],
        Black}]]]
```

```
In[·]:= convex[100]
```

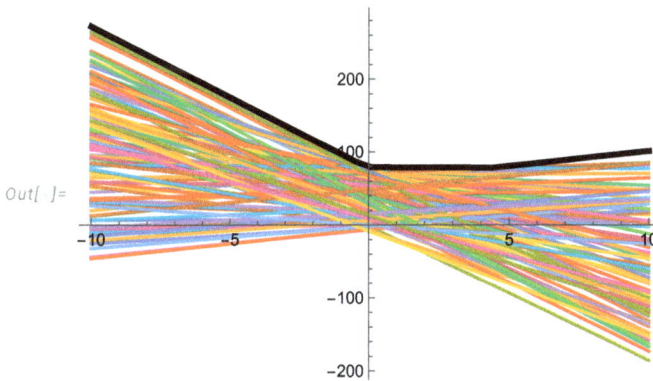

Out[ ]=

Figure 5.21

───────────
3 See Section 1.5 in C. Niculescu, L.-E. Persson, Convex Functions and Their Applications. A Contemporary Approach. CMS Books in Mathematics, Canadian Mathematical Society, Springer, 2006.
4 https://en.wikipedia.org/wiki/Subderivative

5.4.1 Jensen's inequality

One of the most common applications of convexity is in proving inequalities. The first definition of convexity above (Definition 4) leads by an easy inductive argument to one of the most useful inequalities, *Jensen's inequality*.

Theorem 16 (Jensen's inequality). *Let* $f : I \to \mathbb{R}$ *be a convex function,* $x_1, x_2, \ldots, x_n \in I$ *and let* $t_1, t_2, \ldots, t_n \in [0, 1]$ *be such that* $\sum_{i=1}^{n} t_i = 1$. *Then*

$$f\left(\sum_{i=1}^{n} t_i x_i\right) \leq \sum_{i=1}^{n} t_i f(x_i).$$

5.4.1.1 Example

Let us consider the following inequality:

$$1 + \sqrt[3]{e^{2a}}\sqrt[3]{e^b} \leq \sqrt[3]{(1 + e^a)^2}\sqrt[3]{1 + e^b}.$$

To prove it we first rewrite it in the form

$$1 + e^{2a/3}e^{b/3} \leq (1 + e^a)^{2/3}(1 + e^b)^{1/3}.$$

Let us take the natural logarithm of both sides (we can do it, since both sides are positive and the function log is increasing):

$$\log(1 + e^{2a/3}e^{b/3}) \leq \frac{2}{3}\log(1 + e^a) + \frac{1}{3}\log(1 + e^b).$$

Since $2/3 + 1/3 = 1$ and the function $x \mapsto \log(e^x + 1)$ has the second order derivative,

In[·]:= `D[Log[1 + E^x], {x, 2}] // Simplify`

Out[·]:= $\dfrac{E^x}{(1 + E^x)^2}$

which is positive, the function is convex, and we can use Jensen's inequality:

In[·]:= `F[t1 x1 + t2 x2] <= t1 F[x1] + t2 F[x2]/.`
`{x1 -> a, x2 -> b, t1 -> 2/3, t2 -> 1/3,`
`F -> (Log[1 + E^#] &)}`

Out[·]:= $\mathrm{Log}\left[1 + E^{\frac{2a}{3}+\frac{b}{3}}\right] \leq \dfrac{2}{3}\mathrm{Log}\left[1 + E^a\right] + \dfrac{1}{3}\mathrm{Log}\left[1 + E^b\right]$

The right hand side is $\log((e^a + 1)^{2/3}(e^b + 1)^{1/3})$, hence, applying exp to both sides gives the desired inequality.

6 Sequences and series of functions

In this chapter we will consider the problem of approximating functions by polynomials and representing functions by power series. We will discuss the Taylor series and Mathematica®'s function `Series`. Next we consider uniform and almost uniform convergence of function sequences and series and conditions for continuity and differentiability of their limits and sums.

6.1 Power series continued

We will now return to the subject of power series, which we already introduced in Section 3.10.

A *power series* can be viewed as a generalization of a polynomial. Recall that polynomials are just lists of numbers $(a_0, \ldots, a_d)$ together with rules for adding and multiplying any two such lists. It is convenient to write polynomials in the form $a_0 + a_1 x + a_2 x^2 + \cdots + a_n x^n$, where x is called an indeterminate or a variable. Formal power series are defined in exactly the same way, except that we consider infinite rather than just finite sequences. Such a series can be written in the form $a_0 + a_1 x + a_2 x^2 + \cdots + a_n x^n + \cdots$, where x is again a variable. Two such series can also be added and multiplied (the multiplication being given by the Cauchy product).

One important difference between polynomials and formal power series is that a polynomial always defines a function given by substituting numbers for the variable x. However, in the case of formal power series the situation is more complicated for although we can "substitute" a number for x, the number series thus obtained may not be convergent (it is always convergent when we substitute 0). To obtain a function we need to find the set of points where the series is convergent.

Also recall from Section 3.10 that we consider power series of the form $\sum_{n=0}^{\infty} a_n (x - x_0)^n$, where x_0 is called the center of the series. It is not hard to prove that a power series defines a function that is continuous in the interior of its region of convergence. A result by Abel known as *Abel's continuity/limit theorem* [14, Theorem 9.51] states that the function is also left or right continuous at the endpoints of the region of convergence (provided the region of convergence contains those endpoints).

Any polynomial can be rewritten as a polynomial with any given center. Let us demonstrate how this can be done with Mathematica®. Consider the polynomial

```
In[·]:= poly[x_] := x^3 + 3*x^2 - 2*x + 5
```

We want to rewrite it as a polynomial with center at $x_0 = 1$:

```
In[·]:= poly1[x] = Expand[poly[x] /. x -> y + 1] /. y -> x - 1
Out[·]= 7 + 7 (-1 + x) + 6 (-1 + x)^2 + (-1 + x)^3
```

```
In[·]:= Simplify[poly[x] == poly1[x]]
Out[·]= True
```

https://doi.org/10.1515/9783111533063-006

Later on in this chapter we will see how to use the Taylor expansion for the same purpose. For power series the situation is more complicated, since every series with center x_0 is convergent at $x = x_0$, but a series may not be convergent at every point (hence it will not be possible to rewrite it with the center at a point which lies outside of its region of convergence).

6.1.1 Example

A powerful feature of Mathematica® is its ability to express sums of certain power series in terms of built-in analytic functions. Consider the function

$$f(x) = \sum_{n=0}^{\infty} \frac{(x-1)^n}{\sqrt{n+1}}.$$

Let us first find the region of convergence of the power series and then sketch the graph of the function which is defined by this power series. The region of convergence of the power series is $[0, 2)$:

In[·]:= `SumConvergence[(x - 1)^n/Sqrt[n + 1], n]`
Out[·]:= `Abs[-1 + x] < 1 || x == 0`

We can plot the region of convergence as follows:

In[·]:= `reg = Graphics[{Disk[{0, 0}, 0.04], Thickness[0.01],`
    `Line[{{0, 0}, {1.95, 0}}], Red, Disk[{1, 0}, 0.04],`
    `Black, Circle[{2, 0}, 0.04]}]`

Out[·]=

Figure 6.1

Mathematica® can find an exact formula for this series in terms of the special function PolyLog:

In[·]:= `f[x_] = Sum[(x - 1)^n/Sqrt[n + 1], {n, 0, Infinity}]`

Out[·]:= $\dfrac{\text{PolyLog}[\frac{1}{2}, -1 + x]}{-1 + x}$

Let us plot the function over the region of convergence:

In[·]:= `Show[{Plot[f[x], {x, -2, 2}, AxesOrigin ->`
    `{0, 0}], reg}]`

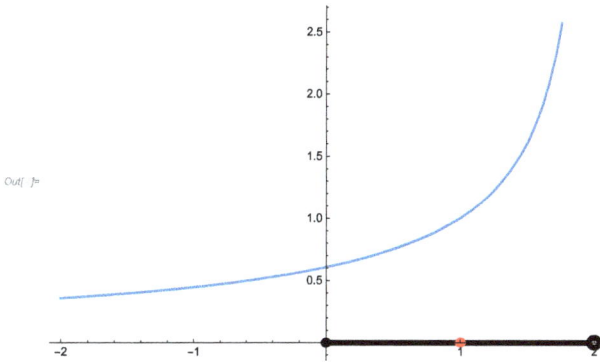

Out[·]=

Figure 6.2

Note that the function itself is defined and differentiable on the entire negative real axis, although the series is not convergent there. It is actually also defined on the positive axis, but it assumes complex values there, e. g.,

In[·]:= N[ComplexExpand[f[3]]]
Out[·]:= -0.805031 - 1.06447 I

6.1.2 Example

Let us consider the series $\sum_{n=0}^{\infty}((-1)^n + 2)^n x^n$. This is an example of a series that Mathematica® cannot deal with in this form and needs human help. Note that applying SumConvergence does not work:

In[·]:= SumConvergence[(2 + (-1)^n)^n*x^n, n]
Out[·]:= SumConvergence[(2 + (-1)^n)^n x^n, n]

Not surprisingly, Mathematica® cannot also find the sum:

In[·]:= Sum[((-1)^n + 2)^n*x^n, {n, 0, Infinity}]
Out[·]:= $\sum_{n=0}^{\infty} (2 + (-1)^n)^n x^n$

Noting that the coefficient of x^n is 1 for n odd and 3^n for n even, we can write this series as the sum of two series $\sum_{n=0}^{\infty} x^{2n+1}$ and $\sum_{n=0}^{\infty} 3^{2n} x^{2n}$ that are easy to deal with. The region of convergence of the original series will then be the intersection of the regions of convergence of these two series.

In[·]:= SumConvergence[3^(2*n)*x^(2*n), n]
Out[·]:= Abs[x] < $\frac{1}{3}$ || 3x == 1 || x == $-\frac{1}{3}$

We have to be careful here, since this answer in Mathematica® 14.2 is clearly wrong, as the series is not convergent when $x = \pm 1/3$.

In[·]:= SumConvergence[x^(2*n + 1), n]
Out[·]:= Abs[x] < 1

Hence the region of convergence is $(-1/3, 1/3)$ and the sum is

In[·]:= f[x_] = Sum[3^(2*n)*x^(2*n), {n, 0, Infinity}] +
        Sum[x^(2*n + 1), {n, 0, Infinity}]

$$Out[·]:= \frac{1}{1 - 9x^2} + \frac{x}{1 - x^2}$$

The graph of the function over its region of convergence is

In[·]:= Plot[f[x], {x, -3^(-1), 1/3}]

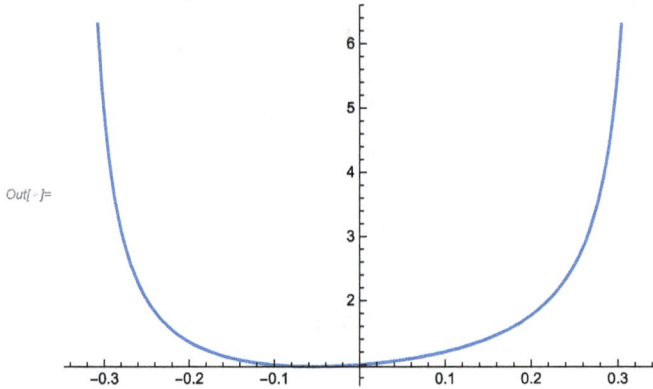

Figure 6.3

Note that again that the function found by Mathematica® is defined on a much larger region:

In[·]:= Plot[f[x], {x, -1, 1}, Exclusions -> {-3^(-1), 1/3}]

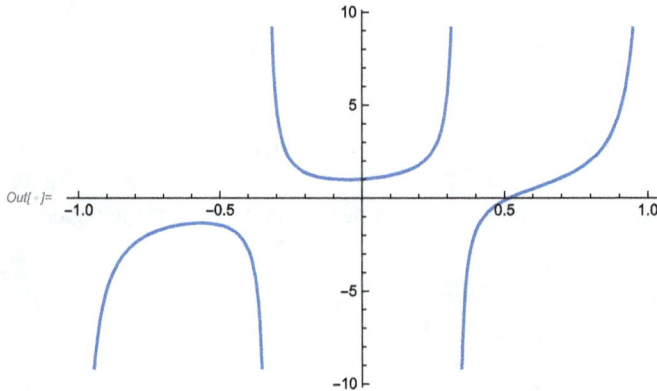

Figure 6.4

As we will see in Section 6.2, although the series that we were given was only defined in the interval $(-1/3, 1/3)$, the function f can also be expressed as a power series outside of this interval, but the series will be a different one with a different center.

Note also that we could find the radius of convergence of the original series by using the *Cauchy–Hadamard Theorem*, which says that the radius of convergence R of a power series $\sum_{n=0}^{\infty} a_n(x - x_0)^n$ is given by

$$\frac{1}{R} = \lim_{n \to \infty} \sup(|a_n|^{1/n}).$$

In Mathematica® 14.2 lim sup is computed by the function MaxLimit:

In[·]:= MaxLimit[Abs[2 + (-1)^n], n -> Infinity]
Out[·]:= 3

Thus the radius of convergence is $R = 1/3$.

6.2 Taylor polynomials and Taylor series

Let $f : D \to \mathbb{R}$ be a function which is n times differentiable at $x_0 \in D$. We define the *n*-th *Taylor polynomial* of f with center at x_0 by $T_n(f) : D \to \mathbb{R}$, where

$$T_n(f)(x) = \sum_{k=0}^{n} \frac{f^{(k)}(x_0)(x - x_0)^k}{k!}.$$

We view $T_n(f)$ as a degree n polynomial approximation to f near x_0. Of course the Taylor polynomial depends on both n and x_0 and sometimes it is denoted by $T_n(f, x_0)$, but for simplicity of notation we will omit the center of expansion.

If the function f is smooth (differentiable infinitely many times), then one can also form its *Taylor series* $\sum_{k=0}^{\infty} f^{(k)}(x_0)(x - x_0)^k/k!$, but it need not converge everywhere where the original function is infinitely differentiable. It may even happen that the Taylor series diverges everywhere except the center [3, Section 24].

The first Taylor polynomial is

$$T_1(f)(x) = f(x_0) + f'(x_0)(x - x_0).$$

Since f is differentiable, we have

$$f(x) - T_1(f)(x) = \left(\frac{f(x) - f(x_0)}{x - x_0} - f'(x_0) \right)(x - x_0).$$

Let

$$r(x) = \frac{f(x) - f(x_0)}{x - x_0} - f'(x_0).$$

By the definition of derivative this can be extended to a function defined on D such that $\lim_{x \to x_0} r(x) = 0$. Hence we have $f(x) = T(f)(x) + R(x)$, where $R(x) = (x - x_0)r(x)$ is some function on D with the property that

$$\lim_{x \to x_0} \frac{R(x)}{x - x_0} = 0.$$

This is generalized to an arbitrary function of class C^n in the following theorem.

Theorem 17 (The Peano Remainder Terms Theorem [1, Section 7.9]). *If $f : D \to \mathbb{R}$ is n times differentiable at $x_0 \in D$ and D contains an open interval I such that $x_0 \in I$ (in such a situation we say that D is a neighborhood of x_0), then*

$$f(x) = T_n(f)(x) + R_n(x),$$

where

$$\lim_{x \to x_0} \frac{R_n(x)}{(x - x_0)^n} = 0.$$

We can also write $R_n(x) = (x - x_0)^n r_n(x)$, where r_n has the property $\lim_{x \to x_0} r_n(x) = 0$.

Any function h that has the property $\lim_{x \to x_0} h(x)/g(x) = 0$ is called an $o(g(x))$ function. Hence the remainder R_n is an $o((x - x_0)^n)$ function. Note that there are a number of useful functions related to asymptotics implemented in Mathematica®.[1]

We can easily show that a representation of f in the form of the sum of a polynomial of degree n and an $o((x - x_0)^n)$ function is unique (we will refer to this representation as *the Peano representation of the function*).

We can use the Peano Remainder Terms Theorem to prove an important result about local maxima and minima (which we mentioned in Chapter 5).

Theorem 18. *Let $f : (a, b) \to \mathbb{R}$ be n times differentiable at $c \in (a, b)$, where $n \geq 2$. Assume that $f^{(k)}(c) = 0$ for $k = 1, 2, \ldots, n-1$ and $a = f^{(n)}(c) \neq 0$. We have*
(i) *If n is even and $a > 0$ ($a < 0$), then f has a strict local minimum (maximum) at c.*
(ii) *If n is odd, then f does not have a local extremum at c.*

In particular, if c is a critical point of f, i.e., $f'(c) = 0$, and if $f''(c) > 0$ ($f''(c) < 0$), then f has a local minimum (maximum) at c.

The proof follows immediately from the Peano Remainder Terms Theorem. We write

$$f(x) - f(c) = \frac{a(x - c)^n}{n!} + (x - c)^n r_n(x), \quad \frac{f(x) - f(c)}{(x - c)^n} = \frac{a}{n!} + r_n(x).$$

If n is even, then the term $(x - c)^n$ is always positive. Since $r_n(x) \to 0$ as $x \to c$, if $a > 0$, then $f(x) > f(c)$ in some neighborhood of c. Hence, there is a strict minimum

---

1 https://reference.wolfram.com/language/guide/Asymptotics.html

at c. Similarly, if $a < 0$, then $f(x) < f(c)$ in some neighborhood of c, hence c is a local maximum. If n is odd, then $f(x)-f(c)$ changes sign when x passes through c, hence there is neither a local maximum nor a local minimum at c.

Given a function f and a point a, we can find the following expansion consisting of the Taylor polynomial of degree n with center at a and the remainder in Mathematica® as follows:

In[·]:= Clear[f]
In[·]:= Series[f[x], {x, a, 2}]
Out[·]:= $f[a] + f'[a] (x - a) + \dfrac{1}{2} f''[a] (x - a)^2 + O[x - a]^3$

Note that the remainder is denoted in Mathematica® by $O[x - a]^{n+1}$, where $n = 2$ in our case. Mathematica®'s O notation does not quite correspond to the mathematical standard "little o" notation. It essentially means that the remainder is of the form $(x - a)^3 h(x)$, where $h(x)$ is some analytic function (that is, one that itself is defined by a power series near a).

There is a convenient short form for Series:

In[·]:= f[x] + O[x, a]^3
Out[·]:= $f[a] + f'[a] (x - a) + \dfrac{1}{2} f''[a] (x - a)^2 + O[x - a]^3$

When a is 0, we can omit it:

In[·]:= f[x] + O[x]^3
Out[·]:= $f[0] + f'[0] x + \dfrac{1}{2} f''[0] x^2 + O[x]^3$

Note that Mathematica® by default assumes that the symbolic functions are analytic and can be expanded as power series:

In[·]:= Series[g[x]*Cos[x], {x, 0, 2}]
Out[·]:= $g[0] + g'[0] x + \dfrac{1}{2} (-g[0] + g''[0]) x^2 + O[x]^3$

It is also possible to tell Mathematica® that the symbolic function is not analytic:

In[·]:= Series[g[x]*Cos[x], {x, 0, 2}, Analytic -> False]
Out[·]:= $g[x] \left(1 - \dfrac{x^2}{2} + O[x]^3\right)$

but we cannot tell Mathematica® that g is of class C^n.

Taylor series with center at 0 are known as *Maclaurin series*. For the function sin we have:

In[·]:= Sin[x] + O[x]^6
Out[·]:= $x - \dfrac{x^3}{6} + \dfrac{x^5}{120} + O[x]^6$

In this case we can obtain the general coefficient of the Taylor series:

$In[\cdot]:=$ SeriesCoefficient[Sin[x], {x, 0, n}]

$$Out[\cdot]:= \begin{cases} \frac{I\,I^n\,(-1+(-1)^n)}{2\,n!} & n \geq 0 \\ 0 & True \end{cases}$$

Note that although the coefficients are real numbers, Mathematica® uses the complex i to obtain a general formula.

The function Series returns a special kind of object and its Head is SeriesData:

$In[\cdot]:=$ FullForm[f[x] + O[x]^2]

$Out[\cdot]//FullForm=$ SeriesData[x, 0, List[f[0],
          Derivative[1][f][0]], 0, 2, 1]

SeriesData returned by the function Series does not represent an infinite series but only a formula essentially equivalent to the Peano representation of a function. Such an object always contains information about a finite number of series coefficients. For example, if

$In[\cdot]:=$ ss = Sum[x^n/n!, {n, 0, 4}] + O[x]^5

$$Out[\cdot]:= 1 + x + \frac{x^2}{2} + \frac{x^3}{6} + \frac{x^4}{24} + O[x]^5$$

then

$In[\cdot]:=$ SeriesCoefficient[ss, {x, 0, 4}]

$$Out[\cdot]:= \frac{1}{24}$$

However,

$In[\cdot]:=$ SeriesCoefficient[ss, {x, 0, 5}]

$$Out[\cdot]:= SeriesCoefficient\left[1 + x + \frac{x^2}{2} + \frac{x^3}{6} + \frac{x^4}{24} + O[x]^5, \{x, 0, 5\}\right]$$

To obtain the n-th Taylor polynomial we use the function Normal (this is a multi-purpose function used in different context which returns what is called "the normal form" of various expressions):

$In[\cdot]:=$ Normal[Series[f[x], {x, a, 2}]]

$$Out[\cdot]:= f[a] + (-a + x)\,f'[a] + \frac{1}{2}(-a + x)^2\,f''[a]$$

As we mentioned earlier, to extract a coefficient in a series expansion we need to use the function SeriesCoefficient:

$In[\cdot]:=$ SeriesCoefficient[f[x] + O[x]^4, {x, 0, 3}]

$$Out[\cdot]:= \frac{1}{6}\,f^{(3)}[0]$$

This is the same answer we can get by asking for a suitable coefficient of the Taylor polynomial:

$In[\cdot]:=$ Coefficient[Normal[f[x] + O[x]^4], x^3]

$$Out[\cdot]:= \frac{1}{6}\,f^{(3)}[0]$$

SeriesCoefficient works both on power series (objects with Head SeriesData) and on analytic functions. It will always compute the coefficient of the n-th term for a numerical n, but in many cases it will not be able to return a general formula for a symbolic n:

In[·]:= SeriesCoefficient[Sin[Cos[x] - 1], {x, 0, n}]
Out[·]:= SeriesCoefficient[-Sin[1 - Cos[x]], {x, 0, n}]

However, for $n = 20$ we have

In[·]:= SeriesCoefficient[Sin[Cos[x] - 1], {x, 0, 20}]
$$Out[·]:= -\frac{60657859289}{2432902008176640000}$$

Mathematica® tries to express a convergent series in terms of built-in analytic functions, for example,

In[·]:= Sum[x^n/(2*n + 1)!, {n, 0, Infinity}]
$$Out[·]:= \frac{\text{Sinh}\left[\sqrt{x}\right]}{\sqrt{x}}$$

Only in the case of such kind of expressions the function SeriesCoefficient will return a coefficient of an arbitrary degree (and in some cases also a symbolic one). If Mathematica® cannot express a series in terms of known analytic functions, it is unable to return the coefficients, even when they are obvious, for example,

In[·]:= SeriesCoefficient[Sum[x^n/(n^2)!, {n, 0,
        Infinity}], {x, 0, 10}]
$$Out[·]:= \text{SeriesCoefficient}\left[\sum_{n=0}^{\infty} \frac{x^n}{n^2!}, \{x, 0, 10\}\right]$$

In such cases the Peano representation gives the desired result:

In[·]:= SeriesCoefficient[Sum[x^n/(n^2)!, {n, 0, 4}] +
        O[x]^5, {x, 0, 4}]
$$Out[·]:= \frac{1}{20922789888000}$$

Now, suppose f itself is a polynomial, e. g.,

In[·]:= f[x_] := 1 - 2*x + 3*x^2 + x^3

What will be its Taylor polynomials with center at 0? The answer follows immediately from the uniqueness of the Peano representation mentioned above:

In[·]:= Table[Normal[Evaluate[f[x]] + O[x]^n], {n, 1, 4}]
Out[·]:= {1, 1 - 2 x, 1 - 2 x + 3 x², 1 - 2 x + 3 x² + x³}

The third and higher Taylor polynomials are equal to the polynomial $f(x)$ itself. The lower ones are just the truncations of the polynomial $f(x)$ after the corresponding degree. Now let us consider the Taylor polynomials of f with centers at points other than

0, e. g., at 1. Again by the uniqueness theorem, the third polynomial will be just the same polynomial but rearranged so that the center is at 1:

In[·]:= Normal[f[x] + O[x, 1]^4]
Out[·]:= 3 + 7 (-1 + x) + 6 (-1 + x)^2 + (-1 + x)^3

In[·]:= Expand[%]
Out[·]:= 1 - 2 x + 3 x^2 + x^3

Note that this also gives us another method of rewriting a polynomial as a polynomial centered around any given number a: we simply use the Taylor polynomial of the same degree centered at a.

As we know, series can be added and can also be multiplied by means of the Cauchy product. We can also perform these operations on their Peano representations. For example

In[·]:= g1 = Cos[x] + O[x]^10; g2 = Cos[x] + O[x]^3;
In[·]:= g1*g2
Out[·]:= 1 - x^2 + O[x]^3

For any series which begins with a non-zero constant term we can find its multiplicative inverse by writing either

In[·]:= 1/(Cos[x] + O[x]^4)
Out[·]:= 1 + $\dfrac{x^2}{2}$ + O[x]^4

or

In[·]:= 1/Cos[x] + O[x]^4
Out[·]:= 1 + $\dfrac{x^2}{2}$ + O[x]^4

Functions which are not analytic at some point can sometimes be expanded in a series centered at that point but this series will not in general be a power series. For example, we can expand

In[·]:= 1/Sin[x] + O[x]^3
Out[·]:= $\dfrac{1}{x}$ + $\dfrac{x}{6}$ + O[x]^3

This kind of series is called a *Laurent series* and it plays an important role in complex analysis but it does not have the properties of power series that will be important for us.

Series (or, more precisely, their Peano representations) can also be composed, just like functions, but only if the center of the first series is the value of the second at its own center. Thus, since cos(0) = 1 we can evaluate

In[·]:= ComposeSeries[Sin[x] + O[x, 1]^4, Cos[x] + O[x]^3]
Out[·]:= Sin[1] - $\dfrac{1}{2}$ Cos[1] x^2 + O[x]^3

However, the following composition

In[·]:= `ComposeSeries[Sin[x] + O[x]^10, Cos[x] + O[x]^10]`

is not defined.

The function `InverseSeries` corresponds to the function `InverseFunction`; in other words, it gives a series for the inverse of the function represented by the original series:

In[·]:= `InverseSeries[Series[Sin[x], {x, 0, 3}]]`

Out[·]:= $x + \dfrac{x^3}{6} + O[x]^4$

In general the function `InverseSeries` may not return a power series but a more general type of series (the *Puiseux series*):

In[·]:= `InverseSeries[1 + x^2 + O[x]^10]`

Out[·]:= $\sqrt{x-1} + O[x-1]^{9/2}$

The main practical application of these operations on series is that they allow us to find the Peano representation of functions, which are compositions of functions whose Peano representations (or Taylor series) are known, without the need to compute derivatives.

Another way to expand a function into a series is by means of `Asymptotic`:

In[·]:= `Asymptotic[Sin[x], {x, 0, Infinity}]`

Out[·]:= $\sum_{n=0}^{\infty} \dfrac{(-1)^n x^{1+2n}}{(1+2n)!}$

In[·]:= `TruncateSum[%, 3]`

Out[·]:= $x - \dfrac{x^3}{6} + \dfrac{x^5}{120}$

This only works if Mathematica® can find an expression for the general term of a series. This works (we omit the large output):

In[·]:= `Asymptotic[Log[Cos[x]], {x, 0, Infinity}];`

but this does not:

In[·]:= `Asymptotic[Exp[Cos[x]], {x, 0, Infinity}]`

Out[·]:= $E^{Cos[x]}$

In the second case one can still obtain finite sums, e. g.,

In[·]:= `Asymptotic[Log[Cos[x]], {x, 0, 5}]`

Out[·]:= $-\dfrac{x^2}{2} - \dfrac{x^4}{12}$

6.2.1 Example

Let us find the limit

$$\lim_{x \to \infty} \left(x - x^2 \log\left(\frac{1}{x} + 1 \right) \right).$$

This problem can be easily solved by using the l'Hospital rule, but we will show how to use the Peano Remainder Terms Theorem. First we convert the problem to one where the limit is taken as the variable goes to 0 (in order to avoid series expansions at infinity). So let us take $y = 1/x$. The problem reduces to finding

$$\lim_{y \to 0} \frac{y - \log(y + 1)}{y^2}.$$

Since

In[·]:= `Log[1 + y] + O[y]^3`

Out[·]:= $y - \dfrac{y^2}{2} + O[y]^3$

we have $\log(1 + y) = y - y^2/2 + r(y)y^2$, where $\lim_{y \to 0} r(y) = 0$. Hence,

$$\lim_{y \to 0} \frac{y - \log(y + 1)}{y^2} = \lim_{y \to 0} \frac{y^2/2 - r(y)y^2}{y^2} = \lim_{y \to 0}\left(\frac{1}{2} - r(y)\right) = \frac{1}{2}.$$

Instead of using the substitution $y = 1/x$ we could have used the Taylor expansion at infinity:

In[·]:= `x - x^2*Log[1/x + 1] + O[x, Infinity]`

Out[·]:= $\dfrac{1}{2} + O\left[\dfrac{1}{x}\right]^1$

This tells us that the function f can be written in the form $1/2 + \tilde{r}(x)$, where $\lim_{x \to \infty} \tilde{r}(x) = 0$. Hence we get the same answer as above.

6.2.2 Example

Let us study for which $a > 0$ the series

$$\sum_{n=1}^{\infty} \left(\frac{1}{\sqrt{n}} - \tan^{-1}\left(\frac{1}{\sqrt{n}}\right)\right)^a$$

is convergent.

The function SumConvergence cannot deal with this problem in Mathematica® 10.4. However in versions 11.3 and 14.2 something surprising happens. The first evaluation of the following expression returns the original input but subsequent evaluation returns the correct solution (provided we do not quit the Kernel in between):

In[·]:= `SumConvergence[(1/Sqrt[n] - ArcTan[1/Sqrt[n]])^a,`
`        n, Assumptions -> {a > 0}]`

Out[·]:= `3 a > 2`

This is probably due to some internal time constraint on the evaluation and the fact that Mathematica® often saves intermediate computation results and reuses them in subsequent evaluations. This would explain why on the second evaluation Mathematica®

has sufficient time to arrive at the solution. We have not been able to test whether performing this computation on a very fast computer would return the answer on the first attempt.

Let us now show how to deal with this problem with the help of the Taylor expansion. First, note that we are dealing with a sum of positive terms:

In[·]:= `Reduce[ArcTan[x] < x, x]`
Out[·]:= $x > 0$

So now let us consider the following Taylor expansion with center at 0 of the function ArcTan:

In[·]:= `ArcTan[x] + O[x]^5`
Out[·]:= $x - \dfrac{x^3}{3} + O[x]^5$

Hence if $f(n) = 1/\sqrt{n} - \tan^{-1}(1/\sqrt{n})$, then $f(n) = 1/(3n^{3/2}) + R_3(n)$, where $\lim_{n\to\infty} R_3(n)/(1/n^{3/2}) = 0$. Hence using the limit comparison test we see that $f(n)$ is similar to $1/n^{3/2}$ and $f(n)^a$ to $1/n^{3a/2}$. Hence the series converges if and only if $a > 2/3$.

6.2.3 Approximating functions by Taylor polynomials

The Peano Remainder Terms Theorem can be viewed as the "local" version of Taylor's theorem, which tells us that the Taylor polynomial of f with center x_0 approximates the function f better and better the closer we are to x_0 (see [11, Chapter 7]). However, this approximation may be good only at x_0. Let us consider the function

In[·]:= `h[x_] := Piecewise[{{0, x == 0}}, Exp[-(x^2)^(-1)]]`
In[·]:= `Plot[h[x], {x, -1, 1}]`

Out[·]=

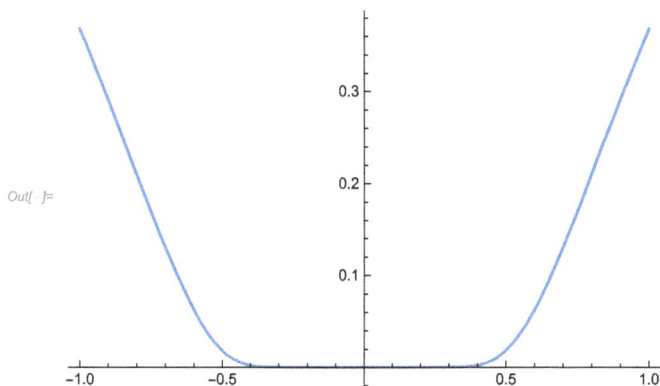

Figure 6.5

This function is differentiable everywhere (including 0) infinitely many times. One can easily show that all of its derivatives at 0 are zero, e. g.,

In[·]:= `Derivative[20][h][0]`
Out[·]:= `0`

Moreover,

In[·]:= `Limit[h[x]/x^n, x -> 0, Assumptions -> n >= 0]`
Out[·]:= `0`

Thus the Taylor formula for h takes the trivial form $h(x) = 0 + h(x)$, where 0 is the n-th Taylor polynomial at zero for every n and h itself is the remainder. In this case the Taylor polynomials with center at zero give us no information about the behavior of the function except at zero. Note that h is an example of an infinitely differentiable function whose Taylor series does not converge to it.

The Peano Remainder Terms Theorem is purely local: it does not help us at all to measure how well a Taylor polynomial approximates the function f in some neighborhood of x. Still, it can be very useful for solving certain problems.

To obtain global information about the approximation of a function by its Taylor polynomials we need a different form for the remainder. There are many of them but the best-known one is Lagrange's form. It requires stronger assumptions on the function f.

Theorem 19 (The Lagrange Remainder Terms Theorem [14, Theorem 8.44]). *Let f be an $(n+1)$ times differentiable function in the interval (a, b) and suppose that the n-th derivative of f is continuous at a and b. Then there exists some $c \in (a, b)$ such that*

$$f(x) = T_n(f)(x) + \frac{f^{(n+1)}(c)}{(n+1)!}(x - x_0)^{n+1}.$$

Note that for $n = 0$ this gives the Mean Value Theorem. In general we know nothing about the value of c except that it lies in the interval (a, b). However, this information is often sufficient to obtain a global estimate of the quality of the Taylor approximation. A very useful consequence is the following statement.

Corollary 1. *Let f be as in Theorem 19 and suppose that $|f^{n+1}(x)| \leq M$ for all $x \in (a, b)$. Then*

$$|f(x) - T_n(f)(x)| \leq \frac{M(b - a)^{n+1}}{(n+1)!}$$

on the whole interval (a, b).

Note that the right hand side tends to 0 as $n \to \infty$. If f is differentiable infinitely many times and if the result above holds for every positive integer n, then the Taylor polynomials are the partial sums of series that converges to f on (a, b). For example, if $f(x) = \sin(x)$, then $f^{(n)}(x)$ is bounded on the entire real line for all n and we see that f is the sum of its Taylor series. Since f was defined as the sum of a convergent power series in Section 3.10 and we know that such a representation must be unique, the Taylor series and the power series centered at 0 which defines the function sin must be the same. Exactly the same is true for the function $f(x) = \exp(x)$. Even though its derivatives are not bounded globally (as in the case of sin or cos) they are still bounded by the same constant for every n on (a, b). The constant now depends on the interval, but in any case, the remainder must tend to 0 as $n \to \infty$ at any point on the real line. More generally, we shall see later on that if a function is defined by a convergent power series centered at x_0 in an interval, then it is infinitely many times differentiable and the series must coincide with the Taylor series at x_0.

The Lagrange form of the remainder contains more information than the Peano one (although it needs stronger assumptions on f) and can therefore be used for the same purposes and for some other purposes for which the Peano form is not suitable.

6.2.4 Example: rational approximation of $\sqrt{e}$

Let us find a rational approximation to $\sqrt{e}$ with error less than 0.001.

This is easy to do in Mathematica®. In fact, the function Rationalize will solve exactly our problem:

In[·]:= Rationalize[Sqrt[E], 0.001]

Out[·]:= $\dfrac{61}{37}$

In[·]:= Abs[% - E^(1/2)] < 0.001
Out[·]:= True

Of course there are many solutions to this problem and Mathematica® chooses the one with the least possible denominator.

Let us now see how this can be done with the help of Taylor polynomials and the Lagrange form of the remainder. Using the Lagrange Remainder Terms Theorem we have

$$e^{1/2} = 1 + \frac{1}{2} + \frac{1}{2!}\left(\frac{1}{2}\right)^2 + \cdots + e^{c_n}\frac{1}{(n+1)!}\left(\frac{1}{2}\right)^{n+1},$$

where c_n is some number between 0 and 1/2, hence it is less than 1/2. Thus the remainder is bounded above by $2(1/2)^{n+1}/(n+1)!$ (since $\sqrt{e} < 2$).

We now use the While loop to find the first n for which this remainder is less than 0.001:

In[·]:= n = 1; While[(2*(1/2)^(n + 1))/(n + 1)! >=
          0.001, n++]; n
Out[·]:= 4

Hence the answer we obtain is

$In[\cdot]:=$ Sum[(1/2)^n/n!, {n, 0, 4}]

$Out[\cdot]:= \dfrac{211}{128}$

$In[\cdot]:=$ Abs[% - E^(1/2)] < 0.001

$Out[\cdot]:=$ True

6.2.5 Example: illustration of approximation of functions with Taylor polynomials

In the interactive illustration below we display the graph of a function and its Taylor polynomials of varying degrees (from 0 to 10). We can manipulate three parameters: the degree of the Taylor polynomial, its center and the region over which we consider the approximation. With the default function sin, we observe that as we increase the degree of the polynomial, the Taylor polynomial approximates the function over a larger and larger region, but the approximation is better near the center of expansion. Our second example is the function $x \mapsto 1/(3 - x)$, which is not defined at $x = 3$ and has two branches. We see that to approximate the branch to the left of the point 3, we need to choose a center to the left of 3, and to approximate the branch to the right, we need to choose a center to the right of 3. The last function is $x \mapsto \log(x + 1)$. For this function the Taylor series with center at 0 is convergent only in the interval $(-1, 1]$ (we only show the graph over the positive axis). To approximate it elsewhere a different center has to be chosen.

```
In[·]:= T[n_, a_: 0][f_] := Normal[f[x] + O[x, a]^(n + 1)]
In[·]:= Manipulate[Show[{Plot[f[x], {x, 0, 2*Pi},
        PlotStyle -> {Thick, Black}, Exclusions ->
        If[Head[f[[1]]] === Power, {3}, None]],
        Plot[Evaluate[T[n, a][f]], {x, a - r, a + r},
        PlotStyle -> Red], Graphics[{Green, PointSize[0.03],
        Point[{a, f[a]}]}]}], {{a, 0, "center"}, 0, 2*Pi,
        Appearance -> "Labeled"}, {{r, 1, "region"}, 0.1,
        2*Pi, Appearance -> "Labeled"}, {{n, 0, "degree"},
        0, 10, 1,  Appearance -> "Labeled"}, {{f, Sin[#1] & ,
        "function"}, {Sin[#1] & , 1/(3 - #1) & ,
        Log[1 + #1] & }, PopupMenu}, SaveDefinitions -> True]
```

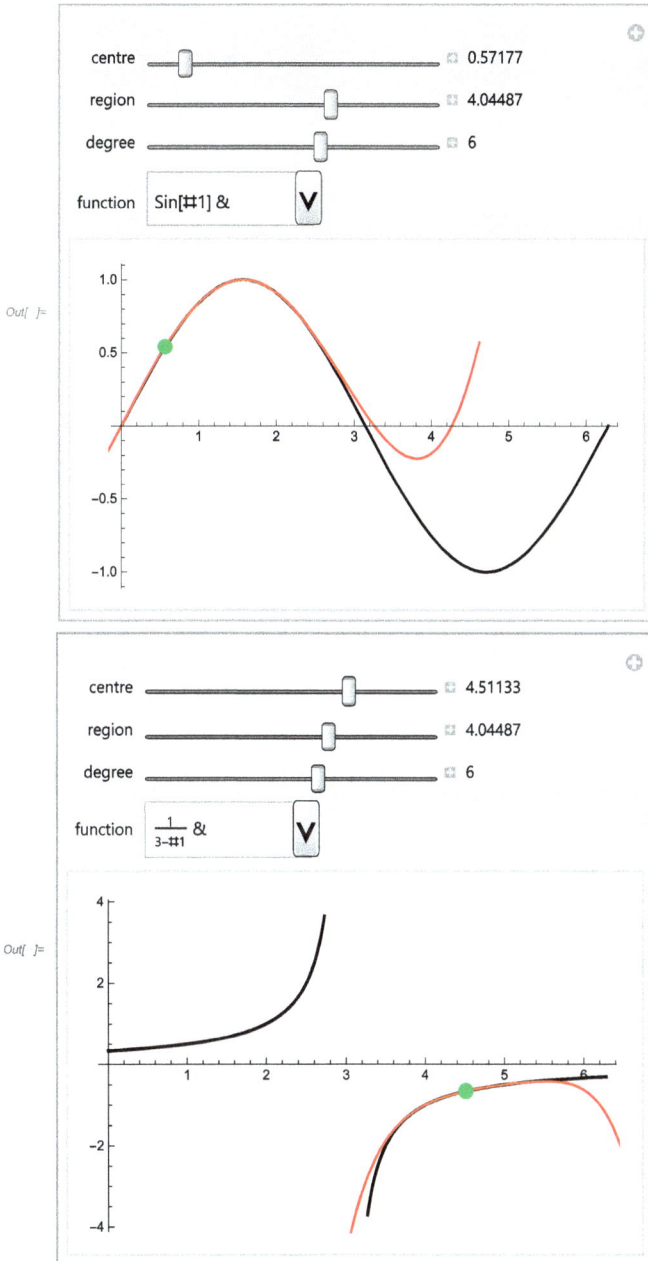

Figure 6.6

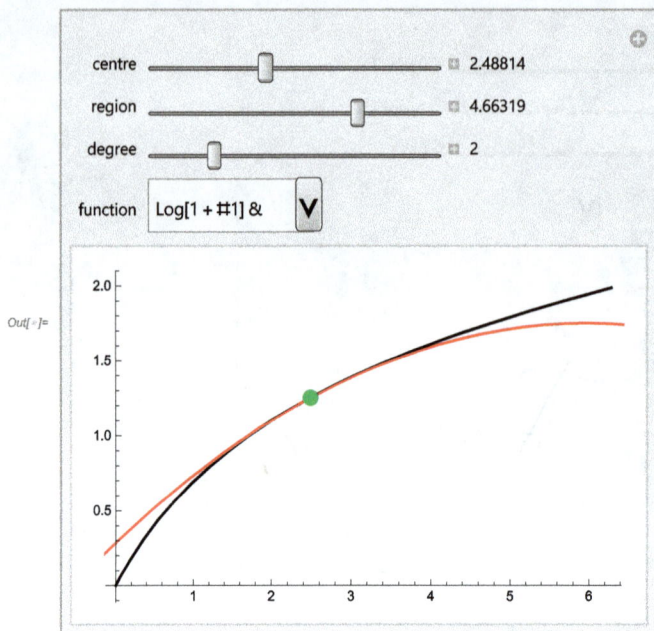

Out[∘]=

Figure 6.6: (continued)

6.2.6 Example

The relation between the properties of an analytic function in some region and the convergence of its Taylor series cannot be fully explained without entering into the field of complex analysis. The following problem is taken from a book of problems in complex analysis[2] but we will solve it entirely using methods of real analysis.

Show that the function

$$f(z) = \sqrt{\frac{z}{1-z}} \arctan\left(\sqrt{\frac{z}{1-z}}\right)$$

is analytic in some neighborhood of 0. Show that it satisfies the differential equation

$$2z(1-z)f'(z) = f(z) + z.$$

By using this differential equation, obtain the coefficients of its Maclaurin series. Finally, show that

$$1 - \frac{2}{3} + \frac{2 \times 4}{3 \times 5} - \frac{2 \times 4 \times 6}{3 \times 5 \times 8} + \cdots = \frac{\log(\sqrt{2}+1)}{\sqrt{2}}.$$

2 J. G. Krzyż, Problems in Complex Variable Theory, American Elsevier Pub. Co., New York, 1971.

We begin by defining the function in Mathematica®:

In[·]:= f[z_] := Sqrt[z/(1 - z)]*ArcTan[Sqrt[z/(1 - z)]]

Let us consider it first as a function on the half-closed interval $[0, 1)$. It is clearly infinitely differentiable on $[0, 1)$. Next, we show that $f(z)$ is everywhere on $[0, 1)$ equal to a power series with center at 0. First, consider the analytic function $u \arctan u$. It is easy to compute its power series. In fact, Mathematica® can give us the n-th coefficient

In[·]:= SeriesCoefficient[u*ArcTan[u], {u, 0, n}]

$$Out[·]:= \begin{cases} -\frac{(-I)^n + I^n)}{2(-1+n)} & n \geq 1 \\ 0 & \text{True} \end{cases}$$

In[·]:= u*ArcTan[u] + O[u]^10

$$Out[·]:= u^2 - \frac{u^4}{3} + \frac{u^6}{5} - \frac{u^8}{7} + O[u]^{10}$$

Alternatively,

In[·]:= ss = Asymptotic[u*ArcTan[u], {u, 0, Infinity}];

In[·]:= TruncateSum[ss, 5]

$$Out[·]:= u^2 - \frac{u^4}{3} + \frac{u^6}{5} - \frac{u^8}{7}$$

It is a power series with only even powers, so we can substitute $u = \sqrt{x}$ and obtain a power series

In[·]:= Sqrt[x]*ArcTan[Sqrt[x]] + O[x]^4

$$Out[·]:= x - \frac{x^2}{3} + \frac{x^3}{5} + O[x]^4$$

By the Cauchy–Hadamard theorem this converges when $|x| < 1$ and hence defines an analytic function on $(-1, 1)$. Now we know that when $|z| < 1/2$, $|z/(1 - z)| < 1$. Indeed, we can show it using Mathematica®, even over the complex numbers:

In[·]:= Reduce[Abs[z] < Abs[z - 1], z, Complexes]

Out[·]:= Re[z] < 1/2

Since $|z| \geq |\operatorname{Re}(z)|$ we see that the function f is analytic at least in the disc $|z| < 1/2$. We now know that it is represented by a power series at least inside this disk. If we could find this power series, we could try to compute its radius of convergence by using the Cauchy–Hadamard theorem. Mathematica® can compute the series numerically

In[·]:= f[z] + O[z]^4

$$Out[·]:= z + \frac{2z^2}{3} + \frac{8z^3}{15} + O[z]^4$$

but it cannot find its general term:

In[·]:= SeriesCoefficient[f[z], {z, 0, n}]

Out[·]:= SeriesCoefficient[Sqrt[z/(1 - z)]*
        ArcTan[Sqrt[z/(1 - z)]], {z, 0, n}]

So we proceed differently. We first prove that f satisfies the differential equation

In[·]:= FullSimplify[2*z*(1 - z)*Derivative[1][f][z]
      == f[z] + z]
Out[·]:= True

We would now like to substitute a power series $\sum_{n=1}^{\infty} a_n z^n$ for $f(z)$ and obtain a recurrence relation between its coefficients. Unfortunately, Mathematica® cannot quite manage this, it can only find a finite number of coefficients, e. g.,

In[·]:= g[z_] := Sum[a[n]*z^n, {n, 1, 5}] + O[z]^4;

In[·]:= SolveAlways[2*z*(1 - z)*Derivative[1][g][z]
      == g[z] + z, z]
Out[·]:= {{a[1] → 1, a[2] → $\frac{2}{3}$, a[3] → $\frac{8}{15}$}}

Alternatively,

In[·]:= AsymptoticDSolveValue[{2*z*(1 - z)*Derivative[1][h][z]
      == h[z] + z, h[0] == 0, Derivative[1][h][0]
      == 1}, h[z], {z, 0, 4}]
Out[·]:= z + $\frac{2z^2}{3}$ + $\frac{8z^3}{15}$ + $\frac{16z^4}{35}$

To deal with the general case, we have to compare the coefficients of z^n of both sides. Setting equal the coefficients of z we obtain $2a_1 = 1 + a_1$, i. e., $a_1 = 1$. For a general n, we get

$$2(n + 1)a_{n+1} - 2na_n = a_{n+1},$$

hence,

$$a_{n+1} = \frac{2n}{2n + 1}a_n.$$

Mathematica® can solve this recurrence relation

In[·]:= a[n] /. RSolve[{a[1] == 1, a[n + 1] ==
      2*(n/(2*n + 1))*a[n]}, a[n], n][[1]]
Out[·]:= Pochhammer[1, -1 + n]/Pochhammer[3/2, -1 + n]

Unfortunately, it expresses the answer in terms of the Pochhammer function, which is difficult to deal with. There is a much easier and obvious solution, in terms of the double factorial

$$a_n = \frac{(2n)!!}{(2n + 1)!!}.$$

Mathematica® can verify that these expressions are the same:

In[·]:= FullSimplify[Pochhammer[1, -1 + n]/
      Pochhammer[3/2, -1 + n] == (2*n - 2)!!/(2*n - 1)!!,
      Assumptions -> Element[n, PositiveIntegers]]
Out[·]:= True

We can now easily show that the radius of convergence is 1. If we have to do it by hand, using the ratio test (d'Alembert's test) is a little easier, but Mathematica® does it equally easily using the Cauchy–Hadamard theorem:

In[·]:= `Limit[((2*n)!!/(2*n + 1)!!)^(1/n), n -> Infinity]`
Out[·]:= `1`

A famous theorem in complex analysis says that the region of convergence of an analytic function in the complex plane is a disc (possibly infinite), on the boundary circle of which there is at least one point where the function becomes infinite. In this case we know that this happens at $z = 1$, but it could possibly happen at some complex number z of modulus less than 1. Of course, the Cauchy–Hadamard theorem assures us that this does not happen, but we can also see it graphically using the function `ComplexPlot3D`. It plots the modulus of $f(z)$ over the complex plane, and uses color to display the value of the argument.

In[·]:= `ComplexPlot3D[Sqrt[z/(1 - z)]*ArcTan[`
`    Sqrt[z/(1 - z)]], {z, 2}, PlotLegends -> Automatic,`
`    AxesLabel -> {"Real", "Imaginary"}]`

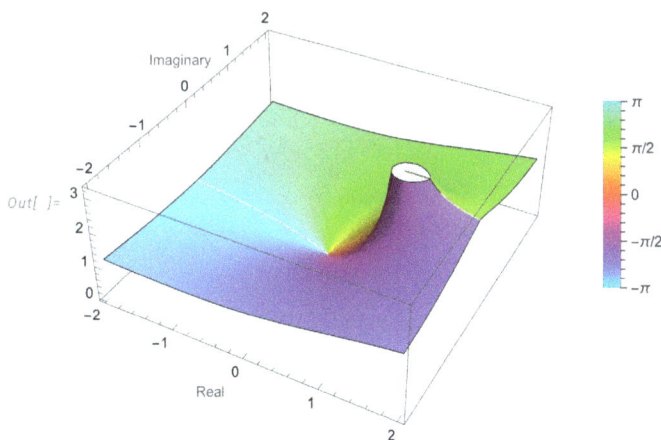

Figure 6.7

Note that both the modulus and the argument are continuous—the latter can be seen from the agreement of the colors as we go round the origin in the complex plane. By contrast, compare this with the graph of the function $\sqrt{z}$:

In[·]:= `ComplexPlot3D[Sqrt[z], {z, 2}, PlotLegends`
`    -> Automatic, AxesLabel -> {"Real", "Imaginary"}]`

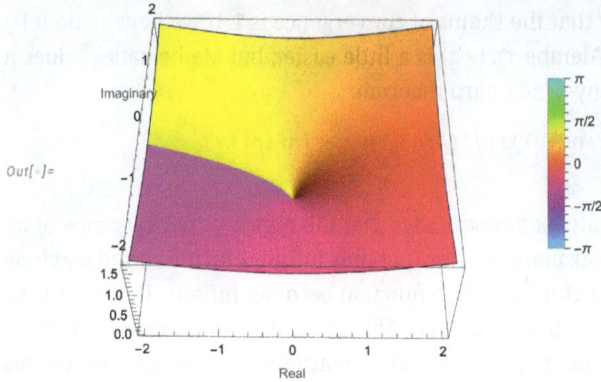

Figure 6.8

Now let us try to see if Mathematica® can find the sum of our power series:

```
In[·]:= Sum[((2*n)!!/(2*n + 1)!!)*z^(n + 1),
        {n, 0, Infinity}]
Out[·]:= (z*ArcSin[Sqrt[z]])/Sqrt[-((-1 + z)*z)]
```

It does find it but the expression it returns looks very different from the one we started from. Let us try to check if they are equal on the interval $(-1, 1)$.

```
In[·]:= h[z_] := (z*ArcSin[Sqrt[z]])/Sqrt[-((-1 + z)*z)];
```

```
In[·]:= FullSimplify[f[z] == h[z], Assumptions -> -1 < z < 1]
Out[·]:= (z*(ArcSin[Sqrt[z]] - ArcTan[Sqrt[
        -(z/(-1 + z))]]))/Sqrt[-((-1 + z)*z)] == 0
```

Mathematica® cannot prove this symbolically. We could try to plot the difference between the two functions but we get a strange-looking result:

```
In[·]:= Plot[f[z] - h[z], {z, -1, 1}]
```

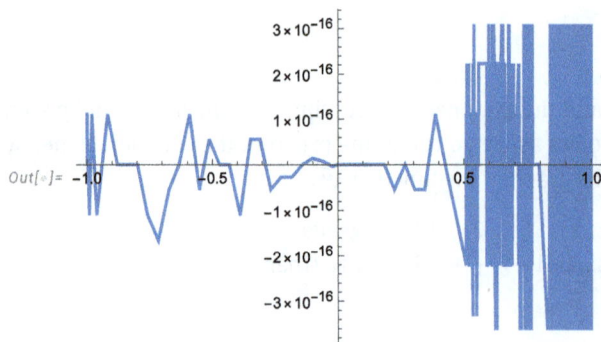

Figure 6.9

It is clear that this is caused by numerical errors which occur when Mathematica® computes with MachinePrecision floating point numbers. We can tell Mathematica® to use
higher precision and the problem disappears:

In[·]:= Plot[f[z] - h[z], {z, -1, 1}, WorkingPrecision -> 20]

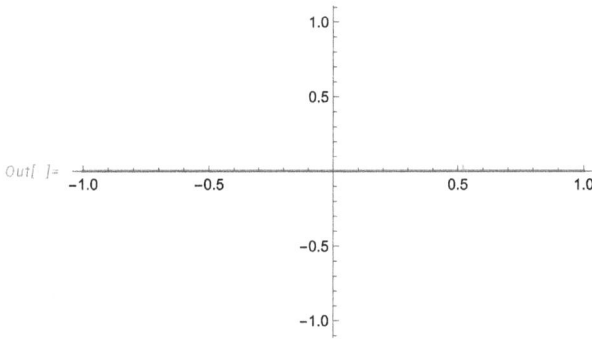

Figure 6.10

Thus we can be quite convinced that the two functions are equal. To give a mathematical
proof that they are, we can show that h also satisfies the same differential equation as f:

In[·]:= FullSimplify[2*z*(1 - z)*Derivative[1][h][z]
        == h[z] + z]
Out[·]:= True

Now proceeding in the same way as for f, we obtain the same power series for h as for
f, hence they have to be equal.

To complete the problem, we observe that the infinite sum

$$1 - \frac{2}{3} + \frac{2 \times 4}{3 \times 5} - \frac{2 \times 4 \times 6}{3 \times 5 \times 8} + \cdots$$

is minus the value of our power series for $z = -1$. Since by the Leibniz criterion this sum
converges, we can use Abel's limit theorem which tells us that its value must be equal to

In[·]:= FullSimplify[-f[-1]]
Out[·]:= ArcCoth[Sqrt[2]]/Sqrt[2]

The same must be true for h, so

In[·]:= TrigToExp[-h[-1]]
Out[·]:= Log[1 + Sqrt[2]]/Sqrt[2]

This is the answer we wanted to get. We can confirm that the two answers are equal:

In[·]:= FullSimplify[ArcCoth[Sqrt[2]]/Sqrt[2] ==
        Log[1 + Sqrt[2]]/Sqrt[2]]
Out[·]:= True

Mathematica® can actually find the sum of the series directly:

```
In[·]:= TrigToExp[Sum[(-1)^n*((2*n)!!/(2*n + 1)!!),
         {n, 0, Infinity}]]
Out[·]:= Log[1 + Sqrt[2]]/Sqrt[2]
```

6.3 Convergence of sequences and series of functions

We started this book with the axioms of the real numbers, which are of course the most basic objects of real analysis. However, we have already pointed out that most (though not all) concepts that we have defined, such as sequences, series and their limits, can also be defined when the real numbers are replaced by complex numbers and most of the theorems have their complex analogues. In fact, much of what we have done so far can be done in a more abstract setting, in which sequences, series, limits, derivatives, etc., are considered in the context of Banach spaces (see [7, Chapter V]). Here we will not do this but only observe that most of the definitions of these concepts and most of the proofs for both real and complex numbers depend on the fact that both real and complex numbers are vector spaces and that they have a notion of "distance", which is given by $|x - y|$, where $|\cdot|$ denotes the absolute value for real numbers and the modulus for complex ones. We will now show that much of the theory of sequences, series and their convergence remains valid if numbers are replaced by functions $f : X \to \mathbb{R}$, where X will be a subset of $\mathbb{R}$. We first need to introduce the concept of "distance" between two functions. If X is compact and the functions are continuous, then the natural concept of distance between two such functions f and g is the maximum distance between their values at the same points in X. For example, consider the functions $f(x) = \cos(x)$ and $g(x) = x - x^3/3$ on the interval $[-1, 1]$:

```
In[·]:= Plot[{Cos[x], x - x^3/3}, {x, -1, 1}, PlotStyle ->
         {Red, Green}, Prolog -> With[{a = -0.636},
         Line[{{a, Cos[a]}, {a, a - a^3/3}}]]]
```

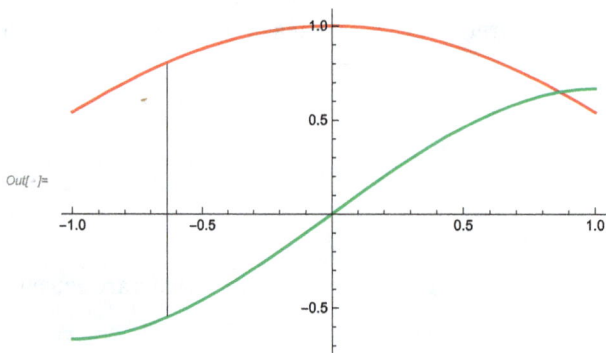

Figure 6.11

What is the distance between these functions? According to our definition it is the largest distance between two points on the red and green curves with the same x-coordinate. It can be computed as follows:

In[·]:= N[Maximize[{Abs[Cos[x] - x + x^3/3], -1 <= x <= 1}, x]]
Out[·]:= {1.35473, {x->-0.636733 + 0. I}}

In[·]:= Chop[%]
Out[·]:= {1.35473, {x -> -0.636733}}

Since functions on X form a vector space, we only need to define the distance between a function and the constant function with value 0. This is called the (supremum) norm of f and is denoted by $\|f\|$:

$$\|f\| = \sup_X |f(x)|.$$

The distance between two functions f and g is then defined by

$$\text{dist}(f, g) = \|f - g\| = \sup_X |f(x) - g(x)|.$$

Note, however, that this will only be a real number if the supremum is attained, i. e., is a maximum. This will always be the case when the functions are continuous and are defined on a compact set. Since distance should always be finite, we will not use this word except when dealing with continuous functions defined on compact sets. However, the norm of a function is defined without these restrictions. The *norm* has the following properties:

(i) $\|f\| \geq 0$.
(ii) $\|f\| = 0$ if and only if $f = 0$ (the zero function).
(iii) If $\lambda \in \mathbb{R}$, then $\|\lambda f\| = |\lambda| \|f\|$.
(iv) $\|f + g\| \leq \|f\| + \|g\|$ (the triangle inequality).

Note that the triangle inequality for the norm implies the "usual" triangle inequality for the distance:

$$\text{dist}(f, g) \leq \text{dist}(f, h) + \text{dist}(h, g).$$

This is true even if we allow distance to take the value ∞.

Now let us consider a sequence $\{f_n\}$ of real-valued functions on some interval (not necessarily compact) $X \subset \mathbb{R}$. We say that such a sequence is *pointwise convergent* if for each $x \in X$ the sequence $\{f_n(x)\}$ is convergent. In this case we can define a function $f : X \to \mathbb{R}$ by $f(x) = \lim_{n \to \infty} f_n(x)$ and we say that the sequence is pointwise convergent to f. As we will soon see, pointwise convergence is not a good notion of convergence for functions as it does not have useful properties. This is because although the values of the functions f_n at a given x approach the value of f, the functions f_n do not become closer to f as $n \to \infty$. The correct notion of convergence for functions is uniform convergence, defined below, which means that $\mathrm{dist}(f, f_n) \to 0$ as $n \to \infty$.

Both pointwise convergence and uniform convergence can be expressed in terms of quantifiers: for *pointwise convergence* we have

$$\forall_{\varepsilon, \varepsilon > 0} \, \forall_{x, x \in X} \, \exists_{N, N \in \mathbb{N}} \, \forall_{n > N} \, |f_n(x) - f(x)| < \varepsilon$$

and for *uniform convergence* we have

$$\forall_{\varepsilon, \varepsilon > 0} \, \exists_{N, N \in \mathbb{N}} \, \forall_{x, x \in X} \quad \forall_{n > N} \, |f_n(x) - f(x)| < \varepsilon.$$

As we can see, just as in the case of ordinary and uniform continuity, the difference between pointwise convergence and uniform convergence can be expressed in terms of the difference in the order of quantifiers. The reader may be interested to know whether there is an example to show uniform convergence using quantifier elimination, just as we did in the case of uniform continuity. Unfortunately this cannot be done because any such statement would involve quantification over the integers and there is no such algorithm (quantifier elimination algorithms exist only over real numbers).

Uniform convergence implies pointwise convergence. The main use of pointwise convergence is that it provides the necessary condition for uniform convergence and in finding the function to which the given sequence could be uniformly convergent. Unfortunately, as we will soon see, uniform convergence on non-compact sets is rather rare; for example, we will see that polynomial sequences do not converge uniformly on $\mathbb{R}$. Fortunately, almost all the important properties of uniform convergence are satisfied for a weaker notion of convergence called *"almost uniform convergence"*. We say that a sequence of functions $\{f_n\}$ converges almost uniformly to f on X if the restrictions of f to all closed intervals contained in X converge uniformly.

6.3.1 Examples: pointwise, uniform and almost uniform convergence of function sequences

(i) Consider the sequence of functions $\{f_n\}$ with $f_n = x^n$ defined on the closed interval $[0, 1]$. Computing $\lim_{n \to \infty} x^n$ we get that the limit is the discontinuous function $f(x) = 0$ for $x < 1$ and $f(1) = 1$, as we see in the following interactive illustration:

```
In[·]:= Manipulate[Plot[Evaluate[{x^n, Piecewise[
          {{0, x < 1}}, 1]}], {x, 0, 1}, PlotStyle ->
          {Black, Orange}, PlotRange -> {0, 1.1},
          Epilog -> {Orange, PointSize[0.02],
          Point[{1, 1}]}, Axes -> False],
          {n, 1, 50, 1, Appearance -> "Labeled"}]
```

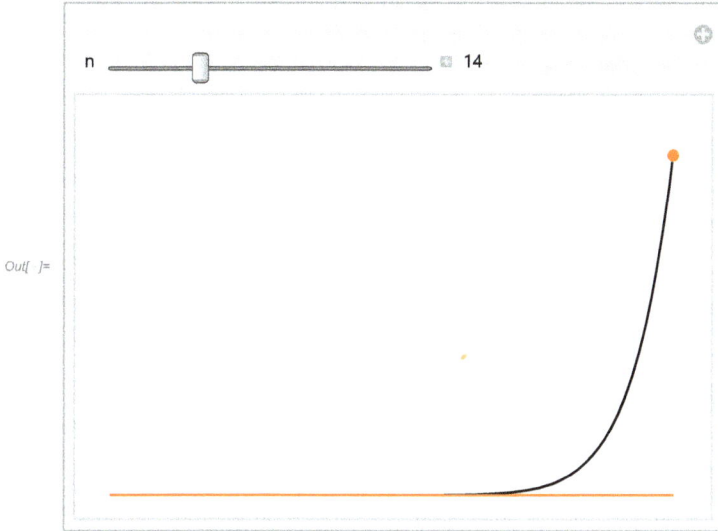

Figure 6.12

On the other hand, for each fixed n, the distance between f_n and f is clearly 1, hence the sequence does not converge uniformly. It also does not converge almost uniformly, since it does not converge on the entire interval [0, 1], which is closed. Note that the limit function is discontinuous although all the functions in the sequence are continuous.

If we restrict the domain to the half closed interval [0, 1) the answer will change. The limit function is now 0 (a continuous functions). The sequence is still not uniformly convergent but it is almost uniformly convergent, since it is clearly uniformly convergent on every closed subinterval of [0, 1).

(ii) Consider the sequence $\{f_n\}$ with $f_n(x) = x^n - x^{n+1}$ on the interval [0, 1], illustrated below:

In[·]:= `Manipulate[Plot[x^n - x^(n + 1), {x, 0, 1},`
`        PlotRange -> {0, 0.25}], {n, 1, 50, 1,`
`        Appearance -> "Labeled"}]`

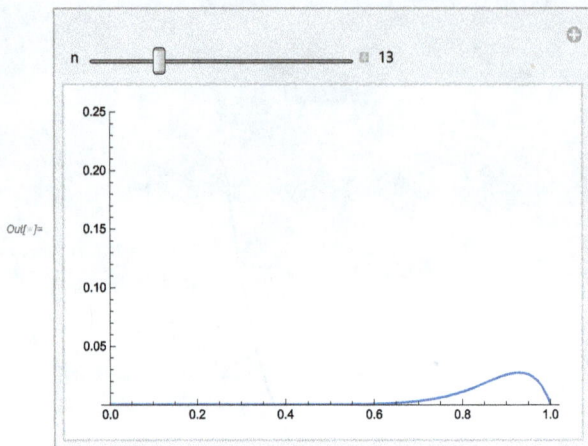

Figure 6.13

The limit function is the constant function 0. To check uniform convergence we compute the distance between the n-th term of the sequence and the limit function, which is just the supremum norm of the n-th term. We first find the critical points:

In[·]:= `Clear[f, n]; f[n_][x_] := x^n - x^(n + 1)`
In[·]:= `x[n_] = x /. First[Solve[D[f[n][x], x] == 0, x]]`
Out[·]:= $\dfrac{n}{1 + n}$

For each n we have only one critical point x_n and since the function is 0 at both ends, f_n assumes the maximum value at this point. We see that

$$\|f_n\| = |f_n(x_n)| = \left(\frac{n}{n+1}\right)^n - \left(\frac{n}{n+1}\right)^{n+1}.$$

Since

In[·]:= `Limit[f[n][x[n]], n -> Infinity]`
Out[·]:= `0`

the sequence $\{f_n\}$ is uniformly convergent.

(iii) For the sequence of functions $\{f_n\}$ with $f_n(x) = x^n - x^{2n}$ the situation is somewhat different. The limit function is again the zero function but a calculation analogous to the one in (ii) gives the value at the moving maximum equal to 1/4, which is independent of n. That means that again the convergence is not uniform on $[0, 1]$ but is almost uniform on $[0, 1)$ since the maximum can be moved to the right of any closed subinterval:

In[·]:= `Manipulate[Plot[x^n - x^(2*n), {x, 0, 1},`
`            PlotRange -> {0, 0.25}], {n, 1, 30, 1,`
`            Appearance -> "Labeled"}]`

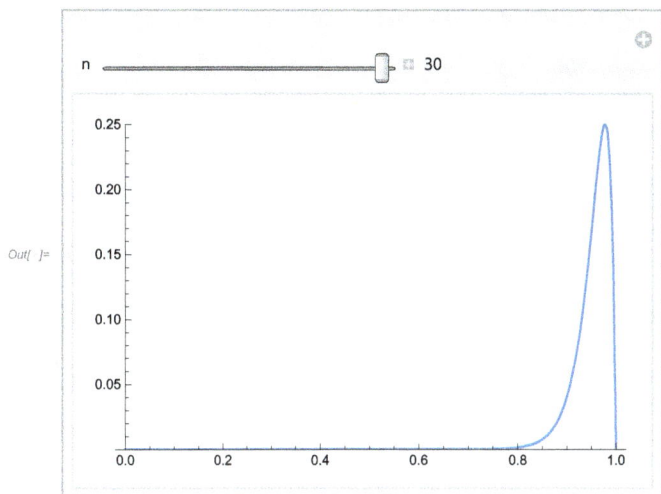

Figure 6.14

(iv) For the sequence $\{f_n\}$ with $f_n(x) = nx/(1 + n + x)$ on $[0, \infty)$ we have

In[·]:= `Limit[(n*x)/(1 + n + x), n -> Infinity,`
`Assumptions -> x > 0]`
Out[·]:= x

So the sequence converges pointwise to the function $f(x) = x$. Is the convergence uniform? The distance between f_n and f is given by

In[·]:= `Maximize[{x - (n*x)/(1 + n + x), x >= 0}, x]`
Out[·]:= `{∞, {x -> Indeterminate}}`

Hence the sequence does not converge uniformly. We can see that if we restrict ourselves to any closed interval (by moving the two locator points in the interactive plot below), then the distance between the two restrictions of the functions in the sequence and the limiting function can be made arbitrarily small:

In[·]:= `Manipulate[Plot[{(n*x)/(1 + n + x), x},`
`{x, 0, 5}, PlotRange -> {{0, 5}, {0, 5}},`
`Prolog -> {Red, Line[{p, q}], Line[`
`{{First[p], (n*First[p])/(1 + n + First[p])},`
`{First[p], First[p]}}], Line[{{First[q],`
`(n*First[q])/(1 + n + First[q])},`
`{First[q], First[q]}}]}], {n, 1, 100, 1,`
`Appearance -> "Labeled"}, {{p, {2, 0}}, {0, 0},`
`{5, 0}, Locator}, {{q, {3, 0}}, {0, 0}, {5, 0},`
`Locator}]`

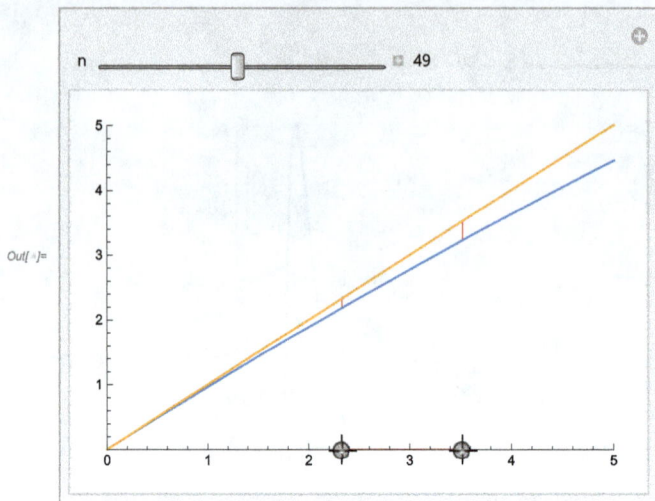

Figure 6.15

6.3.2 Continuity and differentiability of limits and sums

We saw in the first example above (see Section 6.3.1 (i)) that a sequence of continuous functions can converge pointwise to a discontinuous one. One of the most important properties of uniform convergence is the fact that the limit of a uniformly convergent sequence of continuous functions is continuous. This phenomenon is best understood in terms of interchanging limit operations: if $\{f_n\}$ is uniformly convergent on X, then for every $a \in X$ we have the following statement.

Theorem 20 ([14, Theorem 9.12]). *The limit of a uniformly convergent sequence of continuous functions on X is continuous on X. That is, for each $a \in X$*

$$\lim_{x \to a} \left(\lim_{n \to \infty} f_n(x) \right) = \lim_{n \to \infty} \left(\lim_{x \to a} f_n(x) \right).$$

In general, of course, limits cannot be interchanged (one example is the sequence $f_n(x) = x^n$ on $[0, 1]$ at the point 1).

It is obvious that the conclusion of the theorem holds also for sequences of functions which are only almost uniformly convergent, since continuity is a "local property"; in other words, to decide whether a function is continuous or not at some point a we only need to know its values in some neighborhood of a, which can always be taken to be a closed interval. Note that this observation immediately tells us that the sequence in Section 6.3.1 (i) cannot be almost uniformly convergent on $[0, 1]$, because the limit function is not continuous.

The situation with differentiability is more complicated. Let us consider the family of differentiable functions $\{f_n\}$ with $f_n(x) = \sqrt{x^2 + 1/n}$, where $x \in \mathbb{R}$. The sequence is pointwise convergent to

In[·]:= Limit[Sqrt[x^2 + 1/n], n -> Infinity,
          Assumptions -> Element[x, Reals]]
Out[·]:= Abs[x]

which is not differentiable at 0. Evaluating expressions

In[·]:= FullSimplify[Maximize[{Sqrt[x^2 + 1/n] - x,
          x >= 0}, x], Assumptions -> {n > 0}];
In[·]:= FullSimplify[Maximize[{Sqrt[x^2 + 1/n] + x,
          x <= 0}, x], Assumptions -> {n > 0}];

we see that the maximum is $1/\sqrt{n}$ (at $x = 0$), which tends to 0 as $n \to \infty$. Hence, a uniformly convergent sequence of differentiable functions may have a limit that is not differentiable.

In[·]:= Manipulate[Plot[{Sqrt[x^2 + 1/n], Abs[x]},
          {x, -1, 1}, PlotRange -> {{-1, 1}, {0, 2}},
          AxesOrigin -> {0, 0}, AspectRatio -> Automatic],
          {n, 1, 1000, 1, Appearance -> "Labeled"}]

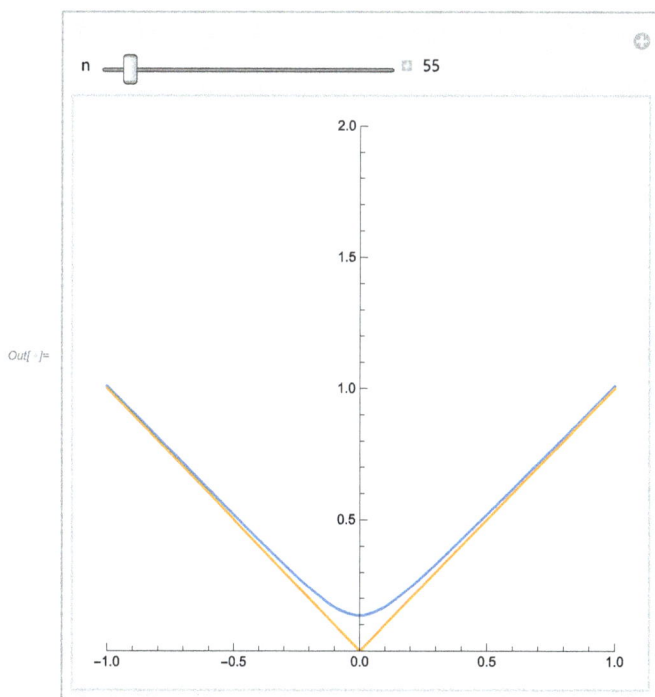

Figure 6.16

Even when a sequence of differentiable functions $\{f_n\}$ converges uniformly to a differentiable function f, it could happen that at some point a the sequence of derivatives $f_n'(a)$ does not converge to $f'(a)$. This happens for the sequence $\{f_n\}$ with $f_n(x) = x/(nx^2 + 1)$ on $\mathbb{R}$, which converges uniformly to 0, but $f_n'(0) = 1$ (see [14, p. 407]):

```
In[·]:= Manipulate[Plot[x/(n*x^2 + 1), {x, -1, 1},
           PlotRange -> {{-1, 1}, {-0.5, 0.5}}],
           {n, 1, 1000, 1, Appearance -> "Labeled"}]
```

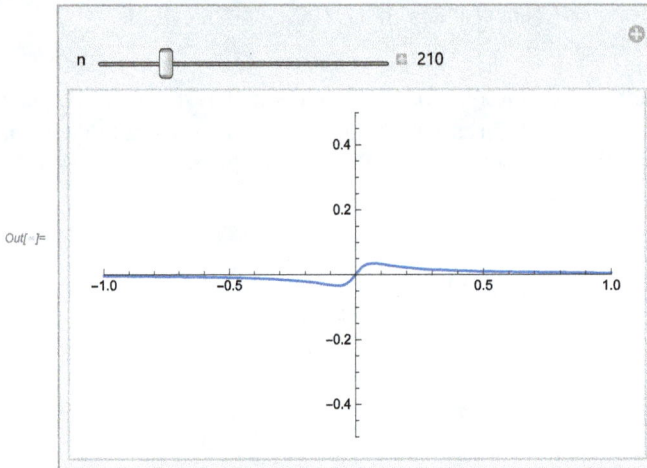

Figure 6.17

A sufficient condition for the limit of a sequence of differentiable functions to be differentiable is given by the following theorem.

Theorem 21 (Differentiation of a sequence of functions [14, Theorem 9.40]). *Suppose that the sequence of functions* $\{f_n\}$ *is such that*
(i) *f_n is of class C^1 on $[a, b]$.*
(ii) *There is some point $x_0 \in [a, b]$ such that the sequence $\{f_n(x_0)\}$ converges.*
(iii) *The sequence of derivatives $\{f_n'\}$ converges uniformly on $[a, b]$ to g.*

Then $\{f_n\}$ converges uniformly to some f on $[a, b]$ such that $f'(x) = g(x)$ on $[a, b]$.

In particular, this means that if a sequence of functions of class C^1 on X is convergent on X and its sequence of derivatives is almost uniformly convergent, then the limit of the sequence is a differentiable function whose derivative is the limit of the derivatives of the functions in the given sequence.

The theory of convergence of sequences of functions is quite analogous to the theory of convergence of number sequences. Most results which can be formulated for number sequences also hold for function sequences with the absolute value $|\cdot|$ replaced by the norm $\|\cdot\|$. For example, the *Cauchy criterion* takes the following form. A sequence of functions $\{f_n\}$ is uniformly convergent if and only if for every $\varepsilon > 0$ there is an integer N such that for every pair of integers n, m larger than N, $\|f_n - f_m\| < \varepsilon$.

Series of functions are defined in the same way as they are defined for numbers: they are simply sequences of partial sums (in which the sums are formed using the usual addition of functions). Thus we can speak of pointwise, uniform and almost uniform convergence of series.

The above theorems about sequences of functions imply corresponding theorems for series. For example, the sum of an almost uniformly convergent series of continuous functions is continuous. Also, if $\{f_n\}$ is a sequence of functions of class C^1 on $[a, b]$ such that for some $x_0 \in [a, b]$ the series $\sum_{n=0}^{\infty} f_n(x_0)$ is convergent and the series $\sum_{n=0}^{\infty} f_n'$ is uniformly convergent on $[a, b]$, then $\sum_{n=0}^{\infty} f_n$ is uniformly convergent on $[a, b]$ to a differentiable function and its derivative is $\sum_{n=0}^{\infty} f_n'$ (see [14, Corollary 9.41]). For this reason, in the case of function series we are almost always interested in determining whether it is (almost) uniformly convergent. Although there is no algorithm that can be used, just as in the case of numbers series, there are a number of tests. Some of them reduce the problem to a problem involving sequences or series of numbers. Next we present some basic tests.

Let $\{f_n\}$ be a sequence of functions defined on X.

(1.) A necessary condition for (almost) uniform convergence of $\sum_{n=0}^{\infty} f_n$ is that the sequence $\{f_n\}$ converges (almost) uniformly to the zero function, which is equivalent to the condition that the sequence of numbers $\{\|f_n\|\}$ converges to 0.

(2.) A sufficient condition for (almost) uniform convergence of $\sum_{n=0}^{\infty} f_n$ is that the series of positive numbers $\sum_{n=0}^{\infty} \|f_n\|$ converges, which follows from the Cauchy criterion. Clearly in this case the series $\sum_{n=0}^{\infty} f_n$ also converges absolutely.

(3.) The most difficult cases are when the necessary condition $\|f_n\| \to 0$ is satisfied but the series $\sum_{n=0}^{\infty} \|f_n\|$ diverges. In such situations we need to use other methods. One of them is the Dirichlet test.

Theorem 22 (Dirichlet's test for uniform convergence [14, Theorem 9.29]). *Suppose that $\{b_n\}$ is a sequence of (non-negative) functions on X such that $b_n(x) \geq b_{n+1}(x)$ and b_n tends to zero uniformly on X. If $\{a_n\}$ is a sequence of functions such that $|s_n(x)| \leq M$ for all n and $x \in X$, where $s_n(x) = \sum_{k=1}^{n} a_k(x)$, then $\sum_{k=1}^{\infty} a_k(x)b_k(x)$ converges uniformly on X.*

The problem with the above sufficient condition (2.) is the need to compute the suprema of f_n over X, which, as we know, can sometimes be a difficult problem. Often, it is easier to find an upper bound for the values of a function than to find the least upper bound (supremum). This makes the following test useful when trying to prove uniform convergence of function series.

Theorem 23 (Weierstrass M-test [14, Theorem 9.25]). *Let $\{M_n\}$ be a sequence of non-negative real numbers and let $\{f_n\}$ be a sequence of functions defined on X, such that $|f_n(x)| \leq M_n$ for all $x \in X$ and each $n \in \mathbb{N}$. If the series $\sum_{k=1}^{\infty} M_k$ converges, then the series $\sum_{k=1}^{\infty} f_k$ converges uniformly (and absolutely) on X.*

Finally we can state a very important result about power series, which follows from the Weierstrass M-test.

Theorem 24 ([14, Theorem 9.26]). *Let $R > 0$ be the radius of convergence of the power series $\sum_{k=0}^{\infty} a_k(x - x_0)^k$ and let $0 < r < R$. Then $\sum_{k=0}^{\infty} a_k(x - x_0)^k$ converges uniformly on $[x_0 - r, x_0 + r]$. In other words, any power series is almost uniformly (and absolutely) convergent in the interior of its region of convergence.*

It follows from this theorem that a power series $\sum_{k=0}^{\infty} a_k(x - x_0)^k$ is infinitely many times differentiable in the interior of its region of convergence and its derivative can be computed by "term-by-term differentiation", i.e., it is equal to the power series $\sum_{k=1}^{\infty} k a_k(x - x_0)^{k-1}$. From this in turn it follows that if a function f can be expressed as a convergent power series, then this series must be its Taylor series. Indeed, if $f(x) = \sum_{k=0}^{\infty} a_k(x - x_0)^k$, then by differentiating repeatedly term-by-term and substituting $x = x_0$ we obtain $a_n = f^{(n)}(x_0)/n!$, which is the n-th coefficient of the Taylor series with center x_0.

We have just seen that a power series determines a continuous function on its region of convergence. This function is differentiable infinitely many times, and all of its derivatives are also given by power series with the same radius of convergence. Note that the derivative of the power series may not be convergent at an endpoint, even if the original power series itself is convergent there:

In[·]:= SumConvergence[x^n*((-1)^n/n), n]
Out[·]:= Abs[x] < 1 || x == 1

but for the derivative we have

In[·]:= SumConvergence[x^n*(-1)^n, n]
Out[·]:= Abs[x] < 1

6.3.3 Examples: pointwise, uniform and almost uniform convergence of function series

(i) Let us investigate pointwise, uniform and almost uniform convergence of the series $\sum_{n=1}^{\infty} x^2/(n^4 + x^4)$ for $x \in \mathbb{R}$.
We easily see that the sequence $\{f_n\}$, where $f_n = x^2/(n^4 + x^4)$, uniformly converges to 0:

In[·]:= Manipulate[Plot[x^2/(n^4 + x^4), {x, -10, 10},
PlotRange -> {0, 1}], {n, 1, 10, 1}]

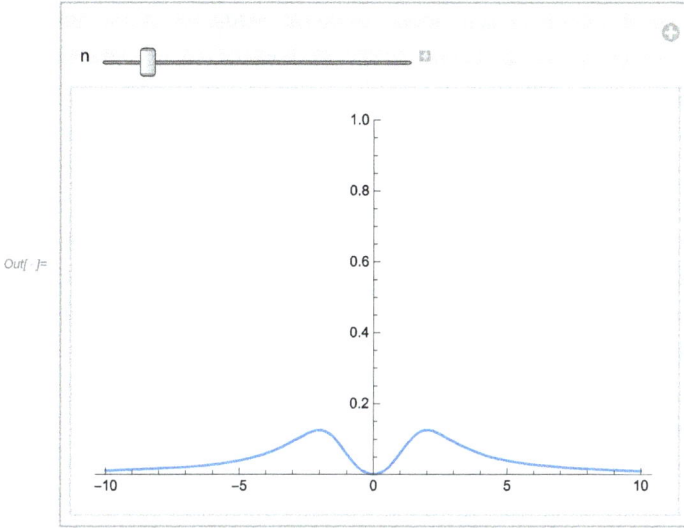

Figure 6.18

In this case we can actually easily find the norms of the functions f_n:

In[·]:= `FullSimplify[Maximize[x^2/(n^4 + x^4), x],`
   `Assumptions -> Element[n, Integers] && n > 0]`

Out[·]:= $\{\dfrac{1}{2\,n^2}, \{x \to -n\}\}$

Since the series $\sum_{n=1}^{\infty} 1/(2n^2)$ converges, our series is uniformly convergent. Instead of computing the norms, we could have observed that the inequality

$$\frac{x^2}{n^4 + x^4} \le \frac{1}{2n^2}$$

holds for all $x \in \mathbb{R}$, hence the result follows from the Weierstrass M-test.

(ii) Let us investigate the convergence of the series $f(x) = \sum_{n=0}^{\infty} xe^{-nx}$ on $(0, \infty)$.
For a fixed x, the series $\sum_{n=0}^{\infty} xe^{-nx}$ is just the geometric series with constant ratio $r = e^{-x}$ with $0 < r < 1$, hence the series is pointwise convergent. Next let us find the norms of the functions $f_n(x) = xe^{-nx}$:

In[·]:= `x /. Solve[D[x/E^(n*x), x] == 0, x, Reals][[1, 1]]`

Out[·]:= $\dfrac{1}{n}$

Hence the maximum of f_n is attained at $x = 1/n$ and is equal to $1/(e\,n)$. Thus, the necessary condition for uniform convergence is satisfied but the series of norms diverges. Hence we have to use a different approach.

In this case the easiest thing to do is to note that we can actually compute the sum of the series s, its partial sums s_n and hence the remainders $r_n = s - s_n$ because they are all sums of geometric series. Thus

$$r_n = s - s_n = \sum_{k=n+1}^{\infty} xe^{-kx} = \frac{xe^{-nx}}{e^x - 1}.$$

So the question of the uniform convergence of our series reduces to the question whether $\{g_n\}$ with $g_n = xe^{-nx}/(e^x - 1)$ converges uniformly to 0 on $(0, \infty)$. This is clearly not true on the whole $(0, \infty)$, but it is true on any compact subset in $(0, \infty)$ either by looking at the following illustration:

In[·]:= `Manipulate[Plot[x/(E^(n*x)*(E^x - 1)), {x, 0, 10},`
   `PlotRange -> {0, 1}], {n, 1, 10, 1}]`

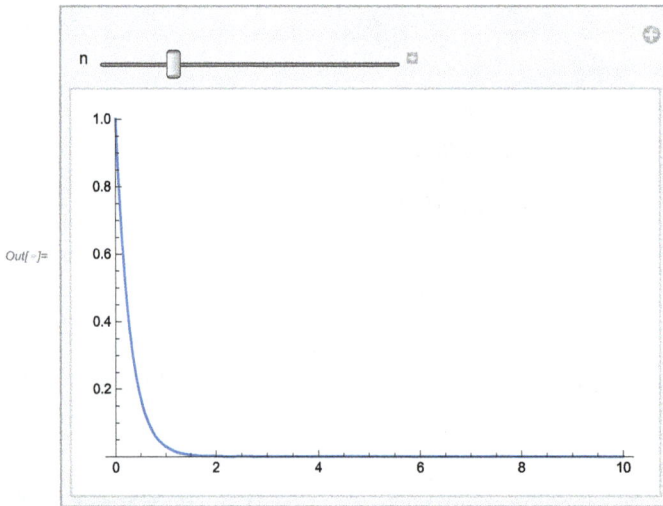

Figure 6.19

or by computing

In[·]:= `Limit[x/(E^(n*x)*(E^x - 1)), x -> 0]`
Out[·]:= 1

Hence the series is not uniformly convergent but is almost uniformly convergent.

6.3.4 Example

Let us show that the function $f(x) = \cos(\sqrt{x})$ is differentiable infinitely many times at 0 and let us find its fifth derivative.

Strictly speaking we should speak of right derivatives or derivatives from above, since the formula is not defined for negative x. Observe, however, that for non-negative x the function $f(x)$ is given by the series $\sum_{n=0}^{\infty}(-1)^n x^n/(2n)!$ and the series is defined and convergent for all x. But, as we said earlier, this means that it must be the Taylor series with center at 0 of the function $g(x) = \sum_{n=0}^{\infty}(-1)^n x^n/(2n)!$, where g is an extension of f to the whole real axis. Hence its fifth derivative at 0 will be

In[·]:= `(5!*(-1)^5)/10!`

Out[·]:= $-\dfrac{1}{30240}$

This can be verified either by a direct computation of derivatives (note that we need to use `Limit` here as direct substitution will not work):

In[·]:= `Limit[D[Cos[Sqrt[x]], {x, 5}], x -> 0]`

Out[·]:= $-\dfrac{1}{30240}$

or by computing the fifth coefficient of the Taylor series:

In[·]:= `5!*SeriesCoefficient[Cos[Sqrt[x]], {x, 0, 5}]`

Out[·]:= $-\dfrac{1}{30240}$

Note that we can actually plot the graph of the function over the negative and positive axis:

In[·]:= `Plot[Cos[Sqrt[x]], {x, -10, 100}]`

Out[ ]=

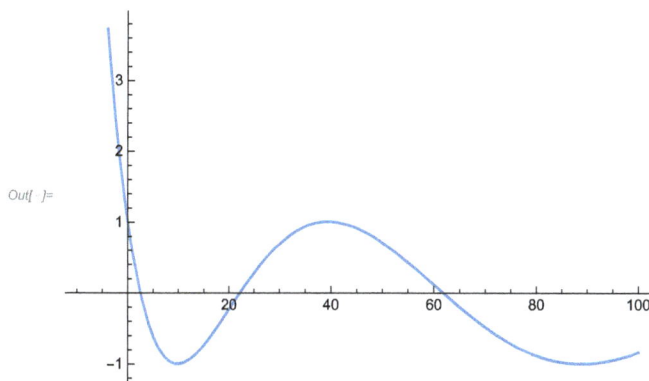

Figure 6.20

The reason why this works is that the functions cos is defined on the whole complex plane and takes real values for pure imaginary arguments, as can be seen by evaluating

In[·]:= `ComplexExpand[Cos[I*x]]`

Out[·]:= `Cosh[x]`

where the function cosh was defined in Section 5.3.3.

Note that

In[·]:= FunctionContinuous[Sin[Sqrt[x]], x]
Out[·]:= False

In[·]:= FunctionContinuous[{Sin[Sqrt[x]], x >= 0}, x]
Out[·]:= True

In[·]:= Plot[Sin[Sqrt[x]], {x, 0, 100}]

Out[·]=

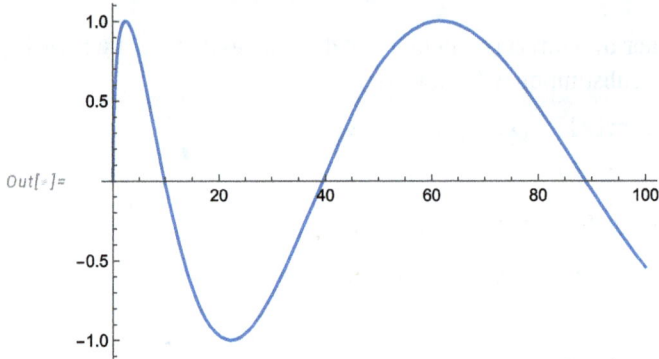

Figure 6.21

7 Integration

In this chapter we first define an indefinite integral (antiderivative) of a function and discuss the Risch algorithm which is used by Mathematica® to compute it. Then we define the Riemann integral, state the fundamental theorem of calculus and consider some applications.

7.1 Indefinite integrals

Let $f : I \to \mathbb{R}$ be a function, where $I \subset \mathbb{R}$ is an interval. We say that a differentiable function $F : I \to \mathbb{R}$ is an *indefinite integral* of f on I (or an *antiderivative* or a *primitive function*) if $F'(x) = f(x)$ for all $x \in I$. This naturally leads to two questions. One is: what kind of functions have antiderivatives? Clearly, not all. For example, we know already that the derivative of any function must have the Darboux property, hence a function that does not have this property, for example, a function that has a "simple jump" discontinuity, cannot have a primitive function.

Later we shall see that any continuous function has an antiderivative (the fundamental theorem of calculus). As we have already seen (example in Section 5.3.2) a function may have an antiderivative even if it is not continuous. Moreover, if a function has an antiderivative, it has infinitely many of them, since if we add any constant to an antiderivative of a function we will get another antiderivative. Note that in practice this means that two antiderivatives of the same function can look quite different: for example both $\sin^2 x$ and $-\cos(2x)/2$ are antiderivatives of $\sin(2x)$. That causes a linguistic difficulty: should one speak of the antiderivative or an antiderivative of a function? We shall usually use the definite article when we refer to the whole class of antiderivatives and to the indefinite one when we want to refer to a particular antiderivative.

The antiderivative of a function f is usually denoted by $\int f(x)\,dx$. In Mathematica® one computes antiderivatives by using the function Integrate, e. g.,

In[·]:= Integrate[x*Cos[x], x]
Out[·]:= Cos[x] + x Sin[x]

In many books on calculus and analysis one writes $x \sin x + \cos x + c$, where c denotes an arbitrary constant. We can obtain this kind of answer by solving the *differential equation*

$$\frac{dF(x)}{dx} = f(x).$$

This is done using the function DSolve. For example:

In[·]:= DSolve[F'[x] == x*Cos[x], F[x], x]
Out[·]:= {{F[x] -> C[1] + Cos[x] + x Sin[x]}}

Integrate by default omits the constant in indefinite integrals. One can add it by using the option GeneratedParameters:

https://doi.org/10.1515/9783111533063-007

In[·]:= Integrate[Exp[x], x, GeneratedParameters -> P]
Out[·]:= E^x + P[1]

Nested integration can be done as follows:

In[·]:= Factor[Integrate[x^a, x, x]]
Out[·]:= x^(2 + a)/((1 + a)*(2 + a))

which is equivalent to

In[·]:= Integrate[Integrate[x^a, x], x]
Out[·]:= x^(2 + a)/((1 + a)*(2 + a))

One can also compute the indefinite integral of a Piecewise function:

In[·]:= f[x_] := Piecewise[{{-x^2, x < 0}, {x^2, True}}]

In[·]:= Integrate[f[x], x]
Out[·]:= Piecewise[{{-(x^3/3), x <= 0}}, x^3/3]

In this case, the derivative of the integral equals the original function:

In[·]:= Simplify[D[%, x] == f[x]]
Out[·]:= True

If one does not want to evaluate the integral, one can use Inactive:

In[·]:= Inactive[Integrate][1/(1 + x), x]
Out[·]:= Inactive[Integrate][1/(1 + x), x]

Differentiating this expression returns the integrand:

In[·]:= D[%, x]
Out[·]:= 1/(1 + x)

Mathematica® can integrate formal expressions, for instance,

In[·]:= Integrate[Derivative[1][f][x], x]
Out[·]:= f[x]

In[·]:= Integrate[((1 - f[x])*Derivative[1][f][x])/E^f[x], x]
Out[·]:= f[x]/E^f[x]

In[·]:= Integrate[Derivative[2][f][x] + 2*a*Derivative[1][f][x], x]
Out[·]:= 2*a*f[x] + Derivative[1][f][x]

The second natural question is: can we find explicitly an antiderivative of a given function? If the function f is continuous, then the fundamental theorem of calculus (which we will discuss later in this chapter) gives us an antiderivative $\int_a^x f(x)\,dx$.

However, one really wishes to have something analogous to the case of differentiation. In other words, we would like to have an algorithm which, for some large family of functions whose antiderivatives belong to this family, computes the antiderivatives of functions obtained by algebraic operations and compositions of functions in the family and expresses them again in terms of functions in the family. Let us make this statement a little more precise. We consider the family of "elementary functions", which includes rational functions, exponentials, logarithms and algebraic functions (e. g., solutions of polynomial equations whose coefficients are elementary functions), as well as trigono-metric, inverse trigonometric, hyperbolic and inverse hyperbolic functions. We require this family to be closed under composition. The question can now be formulated as follows: is there an algorithm which, given an elementary function, returns an elementary function which is its antiderivative? (Of course, this means that when an elementary function does not have an elementary antiderivative the algorithm should inform us of that.)

The study of this problem began in the nineteenth century. For a long time it was believed that no algorithm of this kind could exist. Instead, a number of "heuristic tricks" were developed which try to reduce certain integrals to certain already known ones. One such trick is the so called "integration by parts". This is based on the Leibniz formula for differentiation and takes the form

$$\int f(x)\frac{dg(x)}{dx}\, dx = f(x)g(x) - \int g(x)\frac{df(x)}{dx}\, dx.$$

The point of this kind of formula is that it sometimes allows to replace the problem of finding the antiderivative of a function by the problem of finding the antiderivative of a simpler function. Successful use of such tricks requires skill and luck. For example,

$$\int xe^x\, dx = \int x\frac{de^x}{dx}\, dx = xe^x - \int e^x\frac{dx}{dx}\, dx = e^x(x - 1) + c.$$

However, the indefinite integral $\int xe^{x^2}\, dx$ cannot be found using integration by parts and requires a different trick (substitution). Such a "bag of tricks" is still taught in most university courses. We shall not consider it here, as there are many books on the subject and Mathematica® uses a very different approach. It is based on an algorithm discovered in the 1960s by Risch, based on nineteenth-century work of Liouville and Hermite and early twentiethcentury work of Hardy [8]. The algorithm is known as the *Risch algorithm* and is used by most modern computer algebra systems, including Mathematica®. The algorithm can very rapidly find the antiderivatives of extremely complicated looking functions. For example [8], the following complicated elementary function has an elementary antiderivative:

$$In[\cdot]:= \int ((x\,(x\,+\,1)((x^2\,E^{2x^2} - \text{Log}[x\,+\,1]^2)^2$$
$$+\,2\,E^{3x^2}x(x\,-\,(2x^3\,+\,2x^2\,+\,x\,+\,1)\,\text{Log}[x\,+\,1])))/$$
$$((x\,+\,1)\,\text{Log}[x\,+\,1]^2\,-\,(x^3\,+\,x^2)\,E^{2x^2})^2)\,dx$$

$$Out[\cdot]:= x\,-\,\text{Log}[1\,+\,x]\,-\,\frac{E^{x^2}x\,\text{Log}[1\,+\,x]}{E^{2x^2}x^2\,+\,\text{Log}[1\,+\,x]^2}\,-\,\frac{1}{2}\text{Log}[E^{x^2}x\,-\,\text{Log}[1\,+\,x]]$$
$$+\,\frac{1}{2}\text{Log}[E^{x^2}x\,+\,\text{Log}[1\,+\,x]]$$

In principle the algorithm can decide if the antiderivative of an elementary function can be expressed in an elementary form and returns either this antiderivative or information that no such derivative exists. But this is currently impossible to realize in practice. First of all, the Risch algorithm has many branches and some of them have a very high complexity, which means that a complicated case may take impossibly long to compute. Most computer programs including Mathematica® do not implement such branches at all and when Mathematica® enters into one of them it sometimes returns the answer unevaluated.

There is another problem that results sometimes from the use of the Risch algorithm, illustrated in the following example. Let us consider the rational function

In[·]:= f[x_] := (x^2 + 2*x + 4)/(x^4 - 7*x^2 + 2*x + 17)

This function is defined and continuous on ℝ since its denominator has no roots in ℝ:

In[·]:= Solve[x^4 - 7*x^2 + 2*x + 17 == 0, x, Reals]
Out[·]:= {}

However, the antiderivative returned by Mathematica® is not continuous:

In[·]:= g[x_] = Integrate[f[x], x]

$$Out[·]:= \frac{1}{2} \text{ArcTan}\left[\frac{-1 - x}{-4 + x^2}\right] - \frac{1}{2} \text{ArcTan}\left[\frac{1 + x}{-4 + x^2}\right]$$

In[·]:= Plot[g[x], {x, 1, 3}]

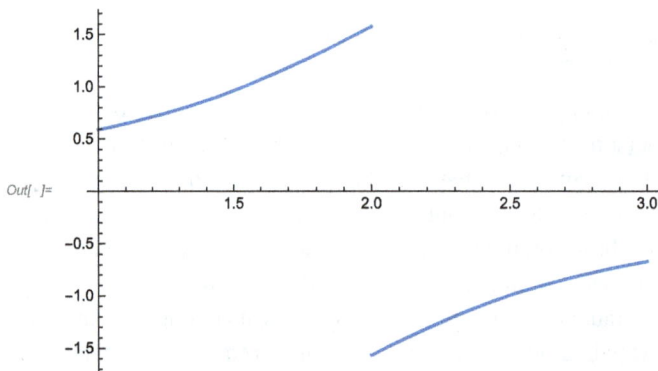

Figure 7.1

As we will see later, when we discuss the fundamental theorem of calculus, we know that a continuous function always has a continuous antiderivative. Looking at the above graph we can see that we could get such an antiderivative by shifting the left branch down or the right branch up so that they meet. We can actually give an explicit formula for a continuous antiderivative by computing first the size of a jump:

In[·]:= Limit[g[x], x -> 2, Direction -> "FromBelow"] -
        Limit[g[x], x -> 2, Direction -> "FromAbove"]
Out[·]:= π

In[·]:= h[x_] := Piecewise[{{g[x], x < 2}, {g[x] + Pi, x >
        2}}, Limit[g[x], x -> 2, Direction -> "FromBelow"]]]
In[·]:= Plot[h[x], {x, 1, 3}]

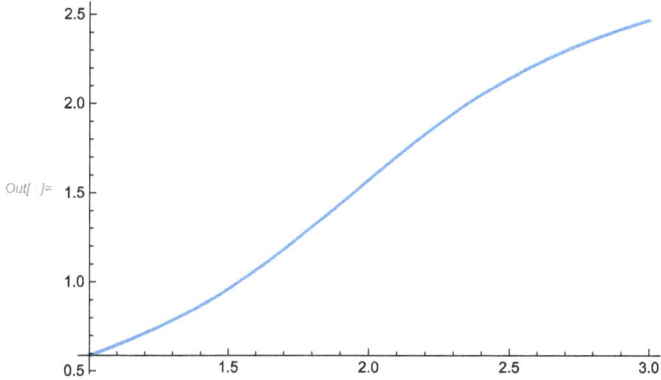

Figure 7.2

The fact that Mathematica® sometimes returns a discontinuous antiderivative of a continuous function is not a bug but a consequence of the way the Risch algorithm works. However, note that when we compute a definite integral over an interval, in which the antiderivative found by the Risch algorithm has a discontinuity, Mathematica® notices that and returns the correct answer:

In[·]:= Integrate[f[x], {x, 1, 3}]
Out[·]:= $\pi - \text{ArcTan}\left[\dfrac{2}{3}\right] - \text{ArcTan}\left[\dfrac{4}{5}\right]$

If we simply used the fundamental theorem of calculus with the discontinuous derivative we would have got an incorrect answer:

In[·]:= g[3] - g[1]
Out[·]:= $-\text{ArcTan}\left[\dfrac{2}{3}\right] - \text{ArcTan}\left[\dfrac{4}{5}\right]$

7.2 The Risch algorithm

We will not try to describe the complete Risch algorithm but we will give a sketch of the main ideas. The most detailed survey can be found in [5, 6].

The first observation is that, although the algorithm finds antiderivatives of both real and complex elementary functions, it actually needs complex numbers to work. When we work over the complex numbers, there are fewer elementary functions since all trigonometric and inverse trigonometric functions can be expressed in terms of the exponential function and its inverse, the logarithm. In Mathematica® we can accomplish this conversion by using the function TrigToExp:

In[·]:= TrigToExp /@ {Sin[x], ArcSin[x], Tan[x], ArcTan[x]}

$$Out[\cdot]:= \left\{ \frac{1}{2} I E^{-Ix} - \frac{1}{2} I E^{Ix}, -I Log \left[I x + \sqrt{1 - x^2} \right], \right.$$

$$\left. \frac{I (E^{-Ix} - E^{Ix})}{E^{-Ix} + E^{Ix}}, \frac{1}{2} I Log[1 - Ix] - \frac{1}{2} I Log[1 + Ix] \right\}$$

There is a problem here: the logarithm as a function on the set of non-zero complex numbers is not an ordinary function but a multi-valued one:

In[·]:= Simplify[Exp[x + 2*Pi*I*n], Element[n, Integers]]
Out[·]:= E^x

Hence, for a complex y the logarithm of y has infinitely many branches differing by integer multiples of $2\pi i$. The Risch algorithm, being purely algebraic, ignores this issue and this sometimes results in the "wrong branch" problem we saw in the example above.

The most important mathematical result which is the basis of the Risch algorithm was actually proved by Liouville in the nineteenth century and is known as *Liouville's Principle*. In order to state it in its modern formulation we need some concepts from modern algebra.

7.2.1 Differential algebras

Definition 5. A *differential field* is a field F together with an operator $\mathcal{D}_F : F \to F$ such that for all $f, g \in F$

$$\mathcal{D}_F(f + g) = \mathcal{D}_F(f) + \mathcal{D}_F(g),$$
$$\mathcal{D}_F(f g) = g \mathcal{D}_F(f) + f \mathcal{D}_F(g).$$

The operator $\mathcal{D}_F$ is called a *derivation* or a *differential operator*.

Definition 6. A differential field $(E, \mathcal{D}_E)$ is called an *extension of a differential field* $(F, \mathcal{D}_F)$ if $F \subset E$ and for all $f \in F$

$$\mathcal{D}_E(f) = \mathcal{D}_F(f).$$

Definition 7. The *field of constants* of $(E, \mathcal{D}_E)$ is the set (actually a field) of elements $K \subset E$ such that

$$K = \{k \in E, \mathcal{D}_E(k) = 0\}.$$

For example, the field of rational functions with rational coefficients $\mathbb{Q}(x) = \{f(x)/g(x)\}$, where f and g are polynomials, is equipped with the differential operator $\mathcal{D}(x) = 1$, which is the ordinary differentiation.

In general $F(\theta)$ denotes the field of rational functions $\{f(\theta)/g(\theta)\}$, where f and g are polynomials with coefficients in F. This is the smallest field that contains F and θ. By induction we can define $F(\theta_1, \theta_2, \ldots, \theta_n) := F(\theta_1, \theta_2, \ldots, \theta_{n-1})(\theta_n)$.

Definition 8. Let F be a differential field and let E be a differential extension of F.

1. An element $v \in E$ for which there exists $u \in F$ such that

$$\mathcal{D}(v) = \frac{\mathcal{D}(u)}{u}$$

   is said to be *logarithmic* over F. In this case we write $v = \log(u)$.
2. An element $v \in E$ for which there exists $u \in F$ such that

$$\frac{\mathcal{D}(v)}{v} = \mathcal{D}(u)$$

   is said to be *exponential* over F. In this case we write $v = \exp(u)$.
3. If for $v \in E$ there exists a polynomial $g \in F[z]$ such that $g(v) = 0$, we say that v is *algebraic* over F. If v is not algebraic over F we say that it is *transcendental* over F.
4. Let E be an extension field of a differential field F. E is called an *elementary extension* of F if it can be obtained from F by successively taking logarithmic, exponential or algebraic extensions. In other words, we can write $E = F(\theta_1, \theta_2, \ldots, \theta_n)$, where θ_i is logarithmic, exponential or algebraic over $F(\theta_1, \theta_2, \ldots, \theta_{i-1})$ for each $i = 1, \ldots, n$.

Now we can state Liouville's Principle.

Theorem 25 (Liouville's Principle). *Let F be a differential field with a constant field K. For $f \in F$, suppose there is some $g \in E$, where E is an elementary extension of F with the same field of constants K, such that $\mathcal{D}(g) = f$. Then there exist $v_0, \ldots, v_m \in F$ and constants $c_1, \ldots, c_m \in K$ such that*

$$f = \mathcal{D}(v_0) + \sum_{i=1}^{m} \frac{c_i \, \mathcal{D}(v_i)}{v_i}.$$

This can also be written as

$$\int f = v_0 + \sum_{i=1}^{m} c_i \log(v_i). \tag{7.10}$$

Liouville's Principle does not, of course, give us an algorithm for finding antiderivatives: such algorithms have to be constructed separately. However, the principle tells us that any elementary function, which has an elementary antiderivative, must have one of the form (7.10) (it may also have antiderivatives which do not have this form). There are, in fact, a number of algorithms which compute antiderivatives in the form (7.10) for different types of extensions. The difference between them concerns essentially efficiency of computation—a very important issue for practical computations but one which we will ignore here.

7.2.2 Example 1: integration of rational functions

Let us first describe the most basic algorithm which illustrates Liouville's Principle—an algorithm for finding antiderivatives of rational functions with coefficients in ℝ. A version of this algorithm is taught in all calculus courses. Rather than trying to describe the general theory, we will consider a typical example. Suppose we wish to find the antiderivative of the following rational function:

In[·]:= f[x_] := (x^5 + x + 1)/(x^4 - 2*x^3 + 2*x^2 - 2*x + 1)

Mathematica® returns the answer

In[·]:= Integrate[f[x], x]

Out[·]:= $\frac{1}{4}\left(-10 - \frac{6}{-1 + x} + 8x + 2x^2 - 4\,\mathrm{ArcTan}[x]\right.$
$\left. + 6\,\mathrm{Log}[-1 + x] + \mathrm{Log}[1 + x^2]\right)$

This does not have the Liouville form (7.10) but we know that we can convert it to one by means of TrigToExp. So let us now consider how to obtain an antiderivative in Liouville's form directly.

In general, given a rational function of the form $p(x)/q(x)$, where the degree of $p(x)$ is larger than that of $q(x)$, we use polynomial division to reduce it to the form $h(x) + p_1(x)/q(x)$, where $\deg p_1(x) < \deg q(x)$. In Mathematica® we can use the function PolynomialQuotientRemainder to do this for us:

In[·]:= pqr = PolynomialQuotientRemainder[Numerator[f[x]],
         Denominator[f[x]], x]
Out[·]:= {2 + x, -1 + 4x - 2x² + 2x³}

Hence $f(x)$ is equal to

In[·]:= First[pqr] + Last[pqr]/Denominator[f[x]]

Out[·]:= $2 + x + \frac{-1 + 4x - 2x^2 + 2x^3}{1 - 2x + 2x^2 - 2x^3 + x^4}$

So we only need to find the antiderivative of

In[·]:= g[x_] := Last[pqr]/Denominator[f[x]]

Next, we compute all the complex roots of the denominator:

In[·]:= roots = x /. Solve[Denominator[g[x]] == 0, x]
Out[·]:= {-I, I, 1, 1}

Thus our denominator factorizes as

In[·]:= Times @@ (x - #1 &) /@ roots
Out[·]:= $(-1 + x)^2\,(-I + x)\,(I + x)$

For this situation when $\deg p(x) < \deg q(x)$, one can prove that an antiderivative of the following form always exists:

$$\int \frac{p(x)}{q(x)} dx = \frac{p_1(x)}{q_1(x)} + \sum_{i=1}^{n} c_i \log(x - a_i),$$

where $q(x) = \prod_{i=1}^{n}(x - a_i)^{d_i}$, $q_1(x) = \prod_{i=1}^{n}(x - a_i)^{d_i-1}$ and $p_1(x)$ is a polynomial with $\deg p_1(x) < \deg q_1(x)$.

In our case an antiderivative of $g(x)$ must have the form

In[·]:= int = a/(x - 1) + b*Log[x - 1] + c*Log[x - I] + d*Log[x + I];

We can now compute the parameters a, b, c, d by

In[·]:= sols = SolveAlways[g[x] == D[int, x], x]

Out[·]:= {{a -> $-\frac{3}{2}$, b -> $\frac{3}{2}$, c -> $\frac{1}{4} + \frac{I}{2}$, d -> $\frac{1}{4} - \frac{I}{2}$ }}

and obtain the antiderivative in Liouville's form:

In[·]:= int /. sols[[1]]

Out[·]:= $-\frac{3}{2(-1 + x)} + \frac{3}{2} Log[-1 + x] + \left(\frac{1}{4} + \frac{I}{2}\right) Log[-I + x] + \left(\frac{1}{4} - \frac{I}{2}\right) Log[I + x]$

If we want to get a solution not containing complex numbers, we can take the real part and use ComplexExpand:

In[·]:= ComplexExpand[Re[%]]

Out[·]:= $-\frac{3}{2(-1 + x)} - \frac{1}{2} Arg[-I + x] + \frac{1}{2} Arg[I + x]$

$+ \frac{3}{4} Log[(-1 + x)^2] + \frac{1}{4} Log[1 + x^2]$

The output still contains I, but the argument function Arg is real-valued and can be expressed in terms of ArcTan. With the help of a little trigonometry we now see that it differs by a constant from the one returned by Mathematica®.

From the point of view of efficiency, the weakest point of the above algorithm is the need to completely factor the denominator of the rational function. There is a method called *Hermite reduction* [5] that instead of complete factorization only requires the so called "squarefree factorization", given in Mathematica® by the function FactorSquareFree:

In[·]:= FactorSquareFree[Denominator[f[x]]]

Out[·]:= $(-1 + x)^2 (1 + x^2)$

See also the Wolfram™ demonstration by S. Blake.[1] Although Hermite reduction does not require the complete factorization, still in general algebraic numbers will appear among the coefficients of the antiderivative. For example:

---

[1] Blake S. Integration using Hermite reduction. Wolfram Demonstrations Project™, published March 7 2011. https://demonstrations.wolfram.com/IntegrationUsingHermiteReduction/

In[·]:= `Integrate[1/(x^3 + x + 1), x]`
Out[·]:= `RootSum[1 + #1 + #1^3 & , Log[x - #1]/(1 + 3*#1^2) & ]`

The function RootSum is just a short notation for the expression which is obtained by applying Normal to it:

In[·]:= `Normal[Integrate[1/(x^3 + x + 1), x]]`

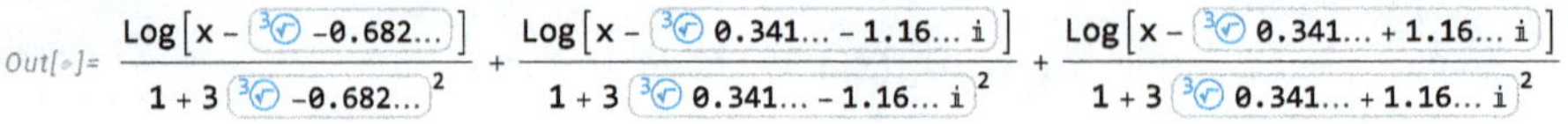

$$Out[\cdot]= \frac{\text{Log}\left[x - \sqrt[3]{\bigcirc}\ -0.682...\right]}{1 + 3\left(\sqrt[3]{\bigcirc}\ -0.682...\right)^2} + \frac{\text{Log}\left[x - \sqrt[3]{\bigcirc}\ 0.341... - 1.16...\ i\right]}{1 + 3\left(\sqrt[3]{\bigcirc}\ 0.341... - 1.16...\ i\right)^2} + \frac{\text{Log}\left[x - \sqrt[3]{\bigcirc}\ 0.341... + 1.16...\ i\right]}{1 + 3\left(\sqrt[3]{\bigcirc}\ 0.341... + 1.16...\ i\right)^2}$$

Figure 7.3

In[·]:= `N[%]`
Out[·]:= `(-0.20861 - 0.18382 I) Log[(-0.34116 - 1.1615 I) +  x]`
` - (0.20861 - 0.18382 I) Log[(-0.34116 + 1.1615 I)`
` + x] + 0.41723 Log[0.68232 + x]`

7.2.3 Example 2: the Risch algorithm for an exponential extension

Let us now consider the situation where we have an exponential extension. Suppose, for example, we want to compute the antiderivative of $1/(e^x + 1)$. Previous versions of Mathematica® returned the answer

In[·]:= `Integrate[1/(E^x + 1), x]`
Out[·]:= `x - Log[1 + E^x]`

Mathematica® 14.1 returns an equivalent expression

In[·]:= `Integrate[1/(E^x + 1), x]`
Out[·]:= `-2*ArcTanh[1 + 2*E^x]`

These two antiderivates differ by a constant

In[·]:= `Simplify[D[%% - %, x]]`
Out[·]:= `0`

Note that the earlier version of Mathematica® returned only exponents and logarithms, and now it returns trigonometric functions and their inverses.

We introduce a new variable $y = e^x$, and we try to compute $\int 1/(y + 1)\, dx$, where $dy/dx = y$. For this situation one can prove that if an antiderivative exists, then it has the following (compatible with Liouville's Principle) form:

$$\int \frac{P(x,y)}{Q(x,y)}\, dx = \frac{p(x,y)}{q(x,y)} + \sum_{i=1}^{m} c_i \log(q_i),$$

where $Q(x, y) = \prod_{i=1}^{m} Q_i^{d_i}(x, y)$, $q(x, y) = \prod_{i=1}^{m} Q_i^{c_i}(x, y)$ with $c_i = d_i - 1$ except when $Q_i(x, y) = y$, in which case $c_i = d_i$, and $q_i(x, y)$ are the irreducible factors of $Q_i(x, y)$, not including y. We write the two-variable polynomial $p(x, y) = \sum_i y^i p_i(x)$, where $p_i(x)$ is a polynomial in x.

Hence in our case we have $Q(x, y) = y + 1$, $Q_1(x, y) = q_1(x, y) = y + 1$, $q(x, y) = 1$. We take $p(x, y) = p_0(x)$ and, hence, the antiderivative must have the form

$$\int \frac{1}{y + 1} \, dx = p_0(x) + c \log(y(x) + 1).$$

Differentiating this we obtain (below we use p0 in Mathematica®'s input and output cells instead of p_0):

In[·]:= D[p0[x] + c*Log[y[x] + 1], x]

Out[·]:= $\dfrac{c\, y'[x]}{1 + y[x]}$ + p0'[x]

Since $y'(x) = y$ we get

In[·]:= int = (c*y)/(y + 1) + p0'[x];

Next we find the unknown function $p_0(x)$ and the coefficient c by using

In[·]:= SolveAlways[int == 1/(y + 1), y]

Out[·]:= {{p0'[x] -> 1, c -> -1}}

(here we should interpret -> as equality). Hence, $p_0(x) = x$, $c = -1$. We finally obtain

In[·]:= int = x - Log[E^x + 1]

If we repeat the same process for the antiderivative $\int x/(e^x + 1) \, dx$, we get a contradiction (because c should be constant). Hence, in this case there is no elementary antiderivative. Indeed, Mathematica® returns the answer

In[·]:= Integrate[x/(E^x + 1), x]

Out[·]:= $\dfrac{x^2}{2}$ - x Log[1 + E^x] - PolyLog[2, -E^x]

which involves the non-elementary function PolyLog.

We will not consider any more examples (the interested reader can consult [9] or the survey article [5] for a complete account or [8]).

7.2.4 Limitations of Mathematica®'s integration

Mathematica® like most other computer algebra systems does not have the full Risch algorithm implemented. For example, let us compute the derivative of an elementary function:

In[·]:= D[x*Sin[x^{ArcSin[x]}, x]

Out[·]:= $x^{1+ArcSin[x]} \cos\left[x^{ArcSin[x]}\right]\left(\dfrac{ArcSin[x]}{x} + \dfrac{Log[x]}{\sqrt{1-x^2}}\right) + Sin\left[x^{ArcSin[x]}\right]$

Now we try to find the antiderivative:

In[·]:= Integrate[%, x]

Out[·]:= $\displaystyle\int\left(x^{1+ArcSin[x]}Cos[x^{ArcSin[x]}]\left(\dfrac{ArcSin[x]}{x} - \dfrac{Log[x]}{\sqrt{1-x^2}}\right) + Sin[x^{ArcSin[x]}]\right)dx$

We see that in this case Mathematica® is unable to find the antiderivative although the fully implemented Risch algorithm should be able to do so.

Note that when expressed in terms of exponential, logarithmic and algebraic functions the integrand takes the form:

In[·]:= TrigToExp[%%]

Out[·]:= $\dfrac{1}{2} I \left(E^{-I x^{-I Log[I x + \sqrt{1-x^2}]}} - E^{I x^{-I Log[I x + \sqrt{1-x^2}]}}\right) + \dfrac{1}{2}\left(E^{-I x^{-I Log[I x + \sqrt{1-x^2}]}} + E^{I x^{-I Log[I x + \sqrt{1-x^2}]}}\right)$
$x^{1-I Log}[I x + \sqrt{1-x^2}]\left(\dfrac{Log[x]}{\sqrt{1-x^2}} - \dfrac{I Log[I x + \sqrt{1-x^2}]}{x}\right)$

It is easy to see that the extension needed to compute this integral is a tower of transcendental and algebraic extensions. The original Risch algorithm did not include this case and it was only worked out by Manuel Bronstein in 1987. The most complete existing implementation of the Risch algorithm was written by Manuel Bronstein and Barry Trager in the computer algebra program Axiom.[2]

Since the Risch algorithm is very complicated and time-consuming, Mathematica® often will not attempt to use it but try to find the derivative in terms of the so called "special functions". "Special functions" is a name used for a large variety of mathematical functions, defined by methods such as power series expansions, solutions to certain differential equations and recursion. Mathematica® can compute the values of such built-in functions to arbitrary precision and knows many identities involving them. Wolfram Research™ has a web site devoted to special functions, http://functions.wolfram. com, where a vast amount of information about them can be found. We have already seen that Mathematica® returns special functions to many integrals for which no elementary representation exists. It also happens sometimes in cases when an elementary antiderivative could be found.

7.3 The Riemann integral

We will only sketch briefly the basic idea of the Riemann integral. For details we refer the reader to [14]. Consider a bounded function f on a closed interval $[a, b]$. A *partition* of $[a, b]$ is a finite ordered set of points $a = x_0 < x_1 < \cdots < x_n = b$, which divides the interval $[a, b]$ into n closed subintervals. Let $\Delta_k = x_k - x_{k-1}$, $k = 1, 2, \ldots, n$.

---

2 https://en.wikipedia.org/wiki/Axiom_(computer_algebra_system)

Let $M_k = \sup_{x \in [x_{k-1}, x_k]} f(x)$, $m_k = \inf_{x \in [x_{k-1}, x_k]} f(x)$. Then the sum $\sum_{k=1}^{n} M_k \Delta_k$ is called an *upper Darboux sum* and $\sum_{k=1}^{n} m_k \Delta_k$ a *lower Darboux sum* of f on $[a, b]$. Geometrically, an upper (lower) sum represents the area of the union of rectangles with $[x_{k-1}, x_k]$ as base and height the maximum (minimum) value of f on $[x_{k-1}, x_k]$. Obviously any lower Darboux sum is smaller than any upper Darboux sum. Note also that when we replace a partition by a subpartition (in other words, we subdivide some intervals) an upper sum will decrease and a lower sum will increase. Let $U(f)$ $(L(f))$ denote the infimum (supremum) of the set of all upper (lower) Darboux sums. This is called the upper (lower) *Darboux integral*. When $U(f) = L(f)$ we say that the function f is *Darboux integrable* and the common value $U(f) = L(f)$ is called the *Darboux (or Riemann) integral* and is denoted by $\int_a^b f(x)\, dx$.

```
In[·]:= f[x_] := Sin[8*x]/2 + 1/2
In[·]:= areaMax[f_, x_][a_, b_] := First[Maximize[{f[x],
          a <= x <= b}, x]]*(b - a)
In[·]:= areaMin[f_, x_][a_, b_] := First[Minimize[{f[x],
          a <= x <= b}, x]]*(b - a)
In[·]:= maX[f_, x_][a_, b_] := First[Maximize[{f[x], a <= x <= b}, x]]
In[·]:= miN[f_, x_][a_, b_] := First[Minimize[{f[x], a <= x <= b}, x]]
In[·]:= Manipulate[Module[{ll, maxs, mins, maxArea, minArea,
          maxRectangles, minRectangles, g2 = Plot[f[x],
          {x, 0, 1}]}, BlockRandom[SeedRandom[rr]; ll =
          Partition[Join[{0}, Sort[RandomReal[{0, 1}, {m}]],
          {1}], 2, 1]; maxs = Apply[maX[f, x], ll, {1}];
          mins = Apply[miN[f, x], ll, {1}];
          maxArea = maxs . Abs[Apply[Subtract, ll, {1}]];
          minArea = mins . Abs[Apply[Subtract, ll, {1}]];
          maxRectangles = Table[Rectangle[{ll[[i, 1]], 0},
          {ll[[i, 2]], maxs[[i]]}], {i, 1, Length[ll]}];
          minRectangles = Table[Rectangle[{ll[[i, 1]], 0},
          {ll[[i, 2]], mins[[i]]}], {i, 1, Length[ll]}];
          Show[Graphics[{Text[StringJoin["max area = ",
          ToString[maxArea]], {0.6, 0.6}], Text[StringJoin[
          "min area = ", ToString[minArea]], {0.6, 0.5}], Red,
          Opacity[0.2], maxRectangles, Blue, minRectangles}],
          g2]]], {{m, 8, "number of points"}, 8, 30, 1,
          ControlType -> PopupMenu}, {{rr, 0, ""},
          Button["new random partition", rr =
          RandomInteger[2^64 - 1]] & }, SaveDefinitions -> True]
```

The illustration above shows upper and lower Darboux sums (pink and violet rectangles) for a continuous function and a random choice of the partition. SeedRandom is used

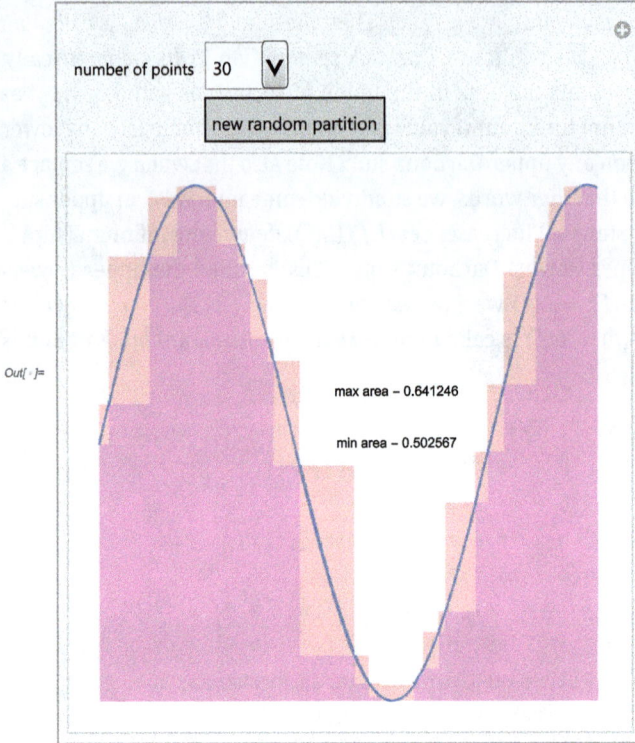

Figure 7.4

to generate a new random sample. BlockRandom assures that only the extra points are generated when we increase the number of points of the partition.

To visualize the area under the function or between two functions one can use

In[·]:= Plot[1/(x^3 + 1), {x, 0, 1}, Filling -> Axis]

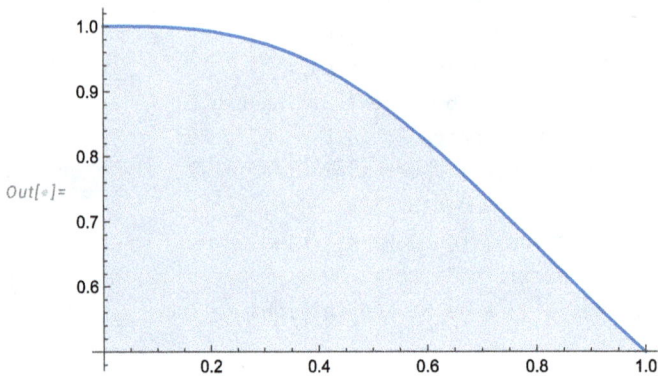

Figure 7.5

or the function `RegionPlot`

In[·]:= `Show[RegionPlot[x^2 < 2*y < 1/(x^4 + 1), {x, -1, 1}, {y, 0, 2}],`
`Plot[{1/2/(1 + x^4), x^2/2}, {x, -2, 2},`
`PlotLegends -> "Expressions"], PlotRange -> {0, 0.8}]`

Out[·]=

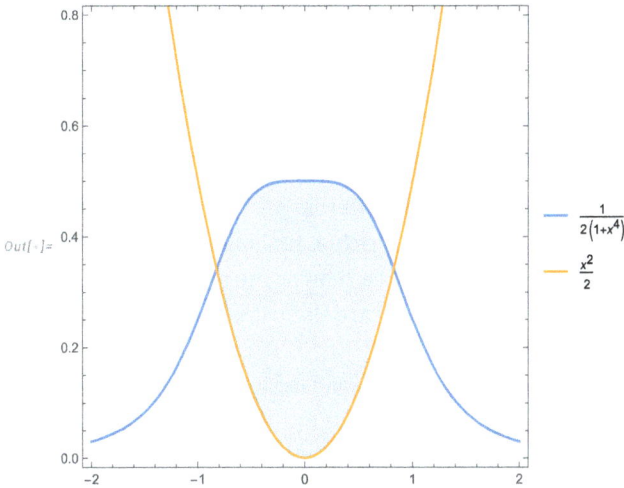

Figure 7.6

Note that if the function is not continuous, then the supremum and infimum values may not be attained on the intervals of the partition. There is an alternative approach to integration due to Riemann. In this approach for each partition $\mathcal{P}$, $a = x_0 < x_1 < \cdots < x_n = b$, we choose points $\bar{x}_1, \ldots, \bar{x}_n$ with $\bar{x}_k \in [x_{k-1}, x_k]$ for $k = 1, 2, \ldots, n$ and consider sums $S_n = \sum_{k=1}^{n} f(\bar{x}_k)\Delta_k$. Clearly this (Riemann) sum always lies between the corresponding Darboux sums. For any $\mathcal{P}$ the norm $\|\mathcal{P}\|$ is defined as the maximum $\max_{1 \leq k \leq n} \Delta_k$. A function f is said to be *Riemann integrable* if and only if there exists a number $\mathcal{I}$ with the following property: for each $\varepsilon > 0$ there is a $\delta > 0$ such that for every Riemann sum σ associated with a partition $\mathcal{P}$ with $\|\mathcal{P}\| < \delta$ one has $|\sigma - \mathcal{I}| < \varepsilon$. The number $\mathcal{I}$ is called the *Riemann integral* of f on $[a, b]$. One can show that $\mathcal{I}$ is independent of the particular choice of partition and subinterval points $\bar{x}_k$ and the concepts of Darboux integral and Riemann integral coincide (see [14, Chapter 6]). From now we shall generally speak only of Riemann integral and Riemann integrability (or just integrability).

The Darboux definition is particularly useful for determining which functions are integrable while the Riemann approach (as we shall soon see) is more useful in concrete computations. With the help of the Darboux definition it is easy to show that all continuous functions and also bounded functions that are continuous except for a countable number of jump discontinuities are integrable (see [14, p. 221] and [14, Theorem 6.21, p. 231]). There are also functions that can be easily shown not to be Riemann integrable, e. g., the Dirichlet function, which is defined by

In[·]:= f[x_] := Piecewise[{{1, Element[x, Rationals]}}]
In[·]:= f /@ {Sqrt[2], Pi, 1/2, E}
Out[·]:= {0, 0, 1, 0}

We immediately see that the function is not Riemann integrable. Indeed, let $x_0, \ldots, x_n$ be any partition of $[0, 1]$. Since every interval contains both rationals and irrationals, we see that $L(f) = 0$ and $U(f) = 1$. The function is thus not Riemann integrable. Nevertheless, Integrate gives

In[·]:= Integrate[f[x], {x, 0, 1}]
Out[·]:= 0

This is, indeed, the correct answer but it needs a different and more powerful concept of integral, the so called Lebesgue integral. The Dirichlet function is not Riemann integrable but it is Lebesgue integrable and Integrate returns the result of the Lebesgue integral. (All functions that are Riemann integrable are also Lebesgue integrable and the corresponding integrals are the same.)

The Darboux definition of the integral can also be used to prove the basic properties of integrals. These follow in a straightforward manner from properties of sums and limits. Here we only state them (for the proofs see [14, Theorems 6.25, 6.26]).

Theorem 26 (General properties of the definite integral). *Let f and g be integrable on $[a, b]$.*

1. *If c_1 and c_2 are constants, then the functions $c_1 f + c_2 g$ are integrable and*

$$\int_a^b (c_1 f(x) + c_2 g(x)) \, dx = c_1 \int_a^b f(x) \, dx + c_2 \int_a^b g(x) \, dx.$$

2. *If $f(x) \le g(x)$ on $[a, b]$, then*

$$\int_a^b f(x) \, dx \le \int_a^b g(x) \, dx.$$

3. *Assume that f is bounded on $[a, b]$ and $c \in [a, b]$. Then f is integrable on $[a, b]$ if and only if f is integrable on $[a, c]$ and $[c, b]$ and*

$$\int_a^b f(x) \, dx = \int_a^c f(x) \, dx + \int_c^b f(x) \, dx.$$

4. *The functions fg and $|f|$ are also integrable on $[a, b]$ and the following inequality holds:*

$$\left| \int_a^b f(x) \, dx \right| \le \int_a^b |f(x)| \, dx.$$

The notions of the definite and indefinite integrals are related by one of the most famous results in mathematics, known as the *Fundamental Theorem of Calculus*. In fact, this theorem is often presented in the form of two theorems, called the First and the Second Fundamental Theorem of Calculus.

Theorem 27 (The First Fundamental Theorem of Calculus, the Newton–Leibniz formula). *If f is integrable on* $[a, b]$ *and F is an antiderivative of f on* $[a, b]$, *then*

$$\int_a^b f(x)\, dx = F(b) - F(a).$$

Note that here we use the word antiderivative in the strict mathematical sense, which means that F has to be a differentiable and therefore continuous function on $[a, b]$. Recall that the Risch algorithm sometimes returns an "antiderivative" that may have points of discontinuity. Nevertheless, Mathematica® generally computes definite integrals correctly (see the example and discussion at the end of Section 7.1).

The First Fundamental Theorem of Calculus can be thought of as saying that integration is a left inverse of differentiation. To see this let us replace b by x and consider $\int_a^x f(t)\, dt$ as a function of x. The theorem now says

$$\int_a^x \frac{dF(t)}{dt} = F(x) - F(a).$$

In other words, integrating the derivative of a function returns the same function (up to a constant). The Second Fundamental Theorem of Calculus says that integration is also the right inverse of differentiation.

Theorem 28 (The Second Fundamental Theorem of Calculus). *Let f be an integrable function on* $[a, b]$. *We define*

$$F(x) = \int_a^x f(t)\, dt$$

for $a \le x \le b$. *Then F is continuous on* $[a, b]$ *and differentiable at every point c at which f is continuous. In this case* $F'(c) = f(c)$.

Note that this gives us a way to construct an antiderivative of any continuous function on $[a, b]$. Mathematica® actually "knows" this theorem:

```
In[·]:= Clear[f]
In[·]:= D[Integrate[f[t], {t, a, x}], x]
Out[·]:= f[x]
```

Note, however, that if we try to do the same thing with a defined function, the evaluation may take much longer and the answer can be complicated:

In[·]:= `f[x_] := 1/(x^3 + 1)`
In[·]:= `D[Integrate[f[t], {t, 1, x}], x]`
Out[·]:= ConditionalExpression[

$$\frac{1}{9}\left(-\frac{3}{(1+\frac{1}{x})x^2} - \frac{3(-1)^{2/3}}{(1-\frac{(-1)^{1/3}}{x})x^2}\right.$$
$$\left. + \frac{3(-1)^{2/3}x(\frac{1}{x} - \frac{(-1)^{2/3}+x}{x^2})}{(-1)^{2/3}+x}\right), \text{Re}[x] > 1 \,\&\&\, \text{Im}[x] == 0]$$

We will explain why we get this kind of answer in the next subsection. Using `Assumptions` makes the answer simpler:

In[·]:= `D[Integrate[f[t], {t, 1, x}, Assumptions -> x > 1], x]`
Out[·]:= $\dfrac{1}{18}\left(\dfrac{6}{1+x} - \dfrac{3(-1+2x)}{1+(-1+x)x} + \dfrac{12}{1+\frac{1}{3}(-1+2x)^2}\right)$

In[·]:= `% // FullSimplify`
Out[·]:= $\dfrac{1}{1+x^3}$

However, if we define f to be a function whose antiderivative Mathematica® does not know, we get the answer quite quickly:

In[·]:= `f[x_] := x^x`
In[·]:= `D[Integrate[f[t], {t, 1, x}], x]`
Out[·]:= x^x

The difference is that in the first case Mathematica® first found a complicated antiderivative using the Risch algorithm and then tried to differentiate it, resulting in a complicated answer (made even more complicated by Mathematica®'s attempt to give conditions for its validity), and in the second case, Mathematica® quickly decided that it could not find an antiderivative so it applied the Second Fundamental Theorem of Calculus to the original input and quickly returned the answer. Hence if we wanted Mathematica® to not evaluate the integral but to use the Second Fundamental Theorem of Calculus, we could do this:

In[·]:= `f[x_] := 1/(x^3 + 1)`
In[·]:= `Block[{f}, D[Integrate[f[t], {t, a, x}], x]]`
Out[·]:= $\dfrac{1}{1+x^3}$

Using `Block` in this way turns f temporarily into an undefined function; Mathematica® then applies the Second Fundamental Theorem and finally replaces the symbol f by its definition.

7.3.1 Using `Integrate` and `NIntegrate` with definite integrals

Here we will try to explain briefly various kinds of definite integrals that appear in Mathematica®. These are obtained by using the function `Integrate` with various kinds of limits (symbolic or numerical) and possibly with assumptions. (There are also numerical integrals that can be computed with `NIntegrate` which we will briefly discuss below.)

The most general type of definite integral in Mathematica® has the form `Integrate[f[x],{x,a,b}]`, where at least one of the limits a and b is symbolic. In this case Mathematica® assumes that the integral is a path integral computed in the complex plane along a straight line joining the points corresponding to the complex numbers a and b. Mathematica® then tries to find the most general complex antiderivative and determine the region of values of a and b in which the formula for the antiderivative is valid. This problem is difficult and Mathematica® generally returns a suboptimal region where the answer is correct. Such formulas can be very complicated. The reader may consider, for example, the following simple definite integral for which Mathematica® gives a very complicated looking answer (which we omit):

In[·]:= `Integrate[1/x, {x, a, b}]`

There are two ways to avoid this issue. We can instruct Mathematica® that we do not want any conditions (which may cause problems when we use the formula):

In[·]:= `Integrate[1/x, {x, a, b}, GenerateConditions -> False]`
Out[·]:= `-Log[a] + Log[b]`

or we can specify conditions on a and b by using `Assumptions`:

In[·]:= `Integrate[1/x, {x, a, b}, Assumptions -> {b > a > 0}]`
Out[·]:= $\mathrm{Log}[\frac{b}{a}]$

When the limits are numerical and exact, an exact numerical answer will be given if Mathematica® is able to compute it:

In[·]:= `Integrate[1/Sqrt[4 - x^2], {x, -1, 1}]`
Out[·]:= $\frac{\pi}{3}$

It is worth mentioning that Mathematica® can integrate various special functions and it knows their integral representations. Moreover, certain integral transforms are also implemented, for instance,

In[·]:= `LaplaceTransform[E^(-t), t, s, GenerateConditions -> True]`
Out[·]:= `ConditionalExpression[1/(1 + s), Re[s] > -1]`

Symbolic integrals can be differentiated with respect to parameters:[3]

---

3 Leibniz integral rule, see https://en.wikipedia.org/wiki/Leibniz_integral_rule

```
In[·]:= D[Integrate[f[x, a], {x, 0, 1}], a]
Out[·]:= Integrate[Derivative[0, 1][f][x, a], {x, 0, 1}]
```

Mathematica® knows how to differentiate with respect to a parameter that appears in both the integrand and endpoints:

```
In[·]:= D[Integrate[f[x, a], {x, p[a], q[a]}], a]
Out[·]:= Integrate[Derivative[0, 1][f][x, a], {x, p[a], q[a]}]
          - f[p[a], a]*Derivative[1][p][a]
          + f[q[a], a]*Derivative[1][q][a]
```

Integrals of vector- and array-valued functions can be computed using lists:

```
In[·]:= Integrate[{2*x, Sin[x]}, {x, 0, 5}]
Out[·]:= {25, 1 - Cos[5]}
```

One can manually perform a change of variables during integration using the function IntegrateChangeVariables. For example, let us apply the change of variables $t = 2x$ to an indefinite integral:

```
In[·]:= IntegrateChangeVariables[Inactive[Integrate]
          [2*Sin[2*x], x], t, t == 2*x]
Out[·]:= Inactive[Integrate][Sin[t], t]
```

Evaluate the result:

```
In[·]:= Activate[%] /. t -> 2*x
Out[·]:= -Cos[2*x]
```

We can compare the result with the original integral:

```
In[·]:= Integrate[2*Sin[2*x], x]
Out[·]:= -Cos[2*x]
```

Another example is a similar change of variables in a definite integral:

```
In[·]:= Activate[IntegrateChangeVariables[Inactive
          [Integrate][2*Sin[2*x], {x, 2, 3}], t, t == 2*x]]
Out[·]:= Cos[4] - Cos[6]
```

One more example is the following:

```
In[·]:= IntegrateChangeVariables[Inactive[NIntegrate]
          [Sqrt[x^2 + y^2], {x, 0, 2}, {y, -Sqrt[4 - x^2], 0}],
          {r, t}, "Cartesian" -> "Polar"]
Out[·]:= Inactive[NIntegrate][r^2, {r, 0, 2}, {t, -Pi/2, 0}]
```

```
In[·]:= Activate[%]
Out[·]:= 4.18879
```

There is NIntegrate, a powerful function that performs numerical integration using many methods (including Riemann sums), controlled by the option Method. This function can compute values of integrals that cannot be found by the exact method of Integrate to arbitrary precision. For example,

In[·]:= Integrate[1/Log[4 + x^3], {x, -1, 1}]

$$Out[·]:= \int_{-1}^{1} \frac{1}{Log[4 + x^3]} dx$$

In[·]:= NIntegrate[1/Log[4 + x^3], {x, -1, 1},
       WorkingPrecision -> 20]
Out[·]:= 1.4547429629288028678

We can also do this:

In[·]:= N[Integrate[1/Log[4 + x^3], {x, -1, 1}]]
Out[·]:= 1.45474

However, the reader should be aware of the fact that this answer was not obtained by Integrate. In fact, Integrate could not find the answer and Mathematica® passed the problem to NIntegrate. In the examples below two different methods are used to obtain the same answer and therefore, they can be used to check the correctness of the answer:

In[·]:= N[Integrate[x^2, {x, -1, 1}]]
Out[·]:= 0.666667

In[·]:= NIntegrate[x^2, {x, -1, 1}]
Out[·]:= 0.666667

NIntegrate is a very reliable function. It is extremely rare for it to return wrong answers, something that happens much more frequently for Integrate.

    Finally, we would like to note that the tutorial in the documentation tutorial/NIntegrateIntroduction gives a lot of information about the function NIntegrate.

7.3.2 Riemann sums

Recall that by a *Riemann sum* of a function f defined on $[a, b]$ we mean a sum of the form $\sum_{k=1}^{n} f(\bar{x}_k)\Delta_k$. The point $\bar{x}_k$ lies in the interval $[x_{k-1}, x_k]$, with the most common choices being $\bar{x}_k = x_{k-1}$, $\bar{x}_k = x_k$ and $\bar{x}_k = (x_{k-1} + x_k)/2$. If f is integrable, then we can compute its integral by choosing a sequence of partitions $\mathcal{P}_k$ with $\|\mathcal{P}_k\| \to 0$ and finding the limit. The most natural "partitioning scheme" is simply division into intervals of equal length, in which case the condition $\|\mathcal{P}_k\| \to 0$ is always satisfied. The integral takes the form $\lim_{n\to\infty} (\sum_{k=1}^{n}(b - a)f(\bar{x}_k)/n)$. Thus this method reduces computing definite integrals to computing limits of Riemann sums. Usually, however, it is easier to compute integrals by using the First Fundamental Theorem of Calculus. In fact we can sometimes reverse the process and compute limits of Riemann sums by finding antiderivatives.

7.3.2.1 Example

Compute $\lim_{n\to\infty}(\sum_{k=n+1}^{2n} 1/k)$.

First note that Mathematica® can compute this limit directly:

In[·]:= `Limit[Sum[1/k, {k, n + 1, 2*n}], n -> Infinity]`
Out[·]:= `Log[2]`

In[·]:= `Sum[1/k, {k, n + 1, 2*n}]`
Out[·]:= `-PolyGamma[0, 1 + n] + PolyGamma[0, 1 + 2 n]`

Mathematica® does this by expressing the sum in terms of special functions and then computing the limit by using its extensive knowledge of such functions. We will find it easier, however, to note that the sum $\sum_{k=n+1}^{2n} 1/k$ can be written in the form

$$\sum_{k=1}^{n} \frac{1}{k+n} = \sum_{k=1}^{n} \frac{1}{n}\frac{1}{1+k/n}.$$

If we divide the interval $[0, 1]$ into n equal parts and take $\bar{x}_k = x_{k-1} = k/n$, we see that we have the n-th Riemann sum of the function $f(x) = 1/(1+x)$ on $[0, 1]$. Hence the limit is

In[·]:= `Integrate[1/(1 + x), {x, 0, 1}]`
Out[·]:= `Log[2]`

7.3.2.2 Example

Sums whose limits as $n \to \infty$ are integrals (Riemann sums) have to have a special form. However, the method of Riemann sums works also for certain sums which are not actually Riemann sums but are "close" to them. Here is one example. Suppose we wish to compute the limit

$$\lim_{n\to\infty} \left(\sum_{k=1}^{n} \frac{n}{8kn + 2k + 5n^2} \right).$$

Mathematica® can do it directly:

In[·]:= `Limit[Sum[n/(8*k*n + 2*k + 5*n^2), {k, 1, n}],`
　　　　`n -> Infinity]`
Out[·]:= $\frac{1}{8}\mathrm{Log}\left[\frac{13}{5}\right]$

Inspecting the summand we see that we are not dealing with a Riemann sum. However, if we remove the summand $2k$ from the denominator, we get a Riemann sum, which can be evaluated by integration:

$$\lim_{n\to\infty} \left(\sum_{k=1}^{n} \frac{n}{8kn + 5n^2} \right) = \lim_{n\to\infty} \left(\frac{1}{n}\sum_{k=1}^{n} \frac{1}{8k/n + 5} \right).$$

This is

In[·]:= Integrate[1/(5 + 8*x), {x, 0, 1}]

Out[·]:= $\dfrac{1}{8}$Log$\left[\dfrac{13}{5}\right]$

We obtain the same answer. Let us try to prove that the two limits are the same without computing them. This is equivalent to

In[·]:= Limit[Sum[n/(8*k*n + 2*k + 5*n^2) -

  n/(8*k*n + 5*n^2), {k, 1, n}], n -> Infinity]

Out[·]:= 0

Consider the sum

$$\sum_{k=1}^{n}\left(\frac{n}{8kn + 2k + 5n^2} - \frac{n}{8kn + 5n^2}\right).$$

The summand is

In[·]:= Together[n/(8*k*n + 2*k + 5*n^2) - n/(8*k*n + 5*n^2)]

Out[·]:= $-\dfrac{2\,k}{(8\,k + 5\,n)(2\,k + 8\,k\,n + 5\,n^2)}$

So we need to show that the limit of the sum

$$\sum_{k=1}^{n}\frac{2k}{(8k + 5n)(8kn + 2k + 5n^2)}$$

is 0 as $n \to \infty$. This sum is clearly smaller than

$$\frac{2n^2}{(5n)(5n^2)} = \frac{2}{25n},$$

which tends to 0 as $n \to \infty$. One can generalize this argument to other sums which are asymptotic Riemann sums.

7.4 Improper integrals

The definite integrals we have considered so far have all involved bounded functions defined on closed intervals. However, in many applications these kinds of integrals are insufficient: sometimes we want to integrate functions which are unbounded or where the domain over which we want to integrate is infinite. Such integrals can be defined by a simple extension of Riemann integration and are known as *improper integrals*. Improper integrals come in two types (kinds), known as the first type and the second type.

7.4.1 Integrals over infinite intervals (improper integrals of the first type)

The first type of integrals are integrals defined over infinite intervals, for example:

In[·]:= Integrate[Log[x]/x^3, {x, 1, Infinity}]

Out[·]:= $\dfrac{1}{4}$

In[·]:= Integrate[Exp[x], {x, -Infinity, 1}]

Out[·]:= E

In[·]:= Integrate[x/E^x^2, {x, -Infinity, Infinity}]

Out[·]:= 0

The first and second integrals above (with only one infinite limit) are defined as limits of ordinary Riemann integrals, e. g.,

$$\int_{1}^{\infty} \frac{\log x}{x^3}\, dx = \lim_{t \to \infty} \int_{1}^{t} \frac{\log x}{x^3}\, dx,$$

$$\int_{-\infty}^{1} e^x\, dx = \lim_{t \to -\infty} \int_{t}^{1} e^x\, dx.$$

They can be computed by Mathematica® by means of these definitions, e. g.,

In[·]:= Limit[Integrate[Log[x]/x^3, {x, 1, t}], t -> Infinity]

Out[·]:= $\dfrac{1}{4}$

but this may take much longer than using infinity directly in the limit, since Mathematica® first attempts to compute an integral with a symbolic limit. Compare, for example,

In[·]:= Integrate[1/(x^3 + 1), {x, 1, Infinity}]

Out[·]:= $\dfrac{1}{9} \left(\sqrt{3}\,\pi - \text{Log}[8] \right)$

with

In[·]:= Limit[Integrate[1/(x^3 + 1), {x, 1, t}], t -> Infinity]

Out[·]:= $Aborted

If we want to use the second method, it is much better to use Assumptions:

In[·]:= Limit[Integrate[1/(x^3 + 1), {x, 1, t},
        Assumptions -> {t > 1}], t -> Infinity]

Out[·]:= $\dfrac{1}{9} \left(\sqrt{3}\,\pi - \text{Log}[8] \right)$

Although generally it is preferable to use the first approach (infinite limits), like in the case of infinite sums, the two methods can return different answers when the integrals are not convergent:

In[·]:= Integrate[1/x, {x, 1, Infinity}]

   ··· Integrate: Integral of $\frac{1}{x}$ does not converge on {1, ∞}.

Out[·]:= $\displaystyle\int_{1}^{\infty} \frac{1}{x} \, dx$

In[·]:= Limit[Integrate[1/x, {x, 1, t}, Assumptions ->
      {t > 1}], t -> Infinity]

Out[·]:= ∞

The following integral is defined as the sum of two integrals:

$$\int_{-\infty}^{\infty} f(x) \, dx = \int_{-\infty}^{c} f(x) \, dx + \int_{c}^{\infty} f(x) \, dx,$$

where c is any number. It is easy to show that the answer is always independent of the choice of c. Thus,

In[·]:= Integrate[x/E^x^2, {x, -Infinity, 1}] +
      Integrate[x/E^x^2, {x, 1, Infinity}]

Out[·]:= 0

There are also situations when the two integrals $\int_{-\infty}^{c} f(x) \, dx$ and $\int_{c}^{\infty} f(x) \, dx$ do not exist, but the symmetric limit $\lim_{t\to\infty} \int_{-t}^{t} f(x) \, dx$ exists. In such cases we call the value of this limit the *Cauchy principal value of the integral* (the integral is still considered divergent). One can compute the Cauchy principal value in Mathematica® using the option PrincipalValue:

In[·]:= Integrate[x^3, {x, -Infinity, Infinity}]

   ··· Integrate: Integral of x³ does not converge on {−∞, ∞}.

Out[·]:= $\displaystyle\int_{-\infty}^{\infty} x^3 \, dx$

In[·]:= Integrate[x^3, {x, -Infinity, Infinity},
      PrincipalValue -> True]

Out[·]:= 0

7.4.2 Improper integrals of the first type and infinite sums

There is both an analogy and a relationship between improper integrals of the first type and infinite sums. All basic convergence tests for series, such as the comparison test, the limit comparison test and the Dirichlet test, have their analogues for improper integrals of the first type (see [14, Theorem 7.5] for the comparison test and [14, Theorem 7.17] for the limit comparison test). The Dirichlet test for integrals is as follows.

Theorem 29 (Dirichlet test for integrals). *If $f, g : [a, \infty) \to \mathbb{R}$, g is decreasing with $\lim_{x\to\infty} g(x) = 0$, for every $r \in [a, \infty)$ f is Riemann integrable on $[a, r]$ and there exists $M \in \mathbb{R}$ such that $|\int_{a}^{r} f(x) \, dx| < M$ for all $r \in [a, \infty)$, then $\int_{a}^{\infty} f(x)g(x) \, dx$ converges.*

For example, this test shows that the integrals $\int_1^\infty (\sin x)/x\, dx$ and $\int_1^\infty (\cos x)/x^{3/2}\, dx$ are convergent. Unlike in the case of sums, Mathematica® does not have a functions for testing convergence of integrals but it performs a test of convergence every time an improper integral is computed:

In[·]:= `Integrate[Sin[x]/x, {x, 1, Infinity}]`
Out[·]:= $\dfrac{1}{2}(\pi - 2\,\mathrm{SinIntegral}[1])$

In[·]:= `Integrate[1/Log[x], {x, 2, Infinity}]`
··· Integrate: Integral of $\frac{1}{\mathrm{Log}[x]}$ does not converge on $\{2,\infty\}$.
Out[·]:= $\displaystyle\int_2^\infty \dfrac{1}{\mathrm{Log}[x]}\, dx$

Note that Mathematica® was able to compute the first integral above in terms of a value of a special function. We can also compute the integral using NIntegrate and compare the answers (as mentioned before they are computed in a different way):

In[·]:= `NumberForm[N[Integrate[Sin[x]/x, {x, 1,`
`        Infinity}], 11], 10]`
Out[·]//NumberForm= `0.6247132564`

In[·]:= `NumberForm[NIntegrate[Sin[x]/x, {x, 1,`
`        Infinity}, PrecisionGoal -> 11,`
`        WorkingPrecision -> 11], 10]`
Out[·]//NumberForm= `0.6247132564`

(NumberForm tells Mathematica® how many digits to show on the screen.)
　　Just like for sums, there is also the concept of absolute convergence for integrals, which is stronger than ordinary convergence. Mathematica® often finds absolute convergence difficult to deal with; for example, it is obvious that $\int_1^\infty |\sin x|/x^2\, dx$ is convergent by the comparison test with the following integral:

In[·]:= `Integrate[1/x^2, {x, 1, Infinity}]`
Out[·]:= `1`

The following computation, however, does not end in a reasonable time:

In[·]:= `Integrate[Abs[Sin[x]]/x^2, {x, 1, Infinity}]`
Out[·]:= `$Aborted`

On the other hand, this is computed quickly:

In[·]:= `Integrate[Sin[x]/x^2, {x, 1, Infinity}]`
Out[·]:= `-CosIntegral[1] + Sin[1]`

　　In addition to the analogy between infinite series and improper integrals of the first type, there is also a more direct relationship.

Theorem 30 (The integral test [14, Theorem 7.25]). *Suppose that f is a non-negative, continuous and decreasing function of x for x ≥ 1. Then $\sum_{k=1}^{\infty} f(k)$ is convergent if and only if $\int_{1}^{\infty} f(x)\,dx$ is convergent. Moreover, when the series converges*

$$\int_{1}^{\infty} f(x)\,dx \leq \sum_{k=1}^{\infty} f(k) \leq \int_{1}^{\infty} f(x)\,dx + f(1)$$

or, equivalently,

$$\sum_{k=2}^{\infty} f(k) \leq \int_{1}^{\infty} f(x)\,dx \leq \sum_{k=1}^{\infty} f(k).$$

A rigorous proof is given in many textbooks (see, for example, [14, p. 300]) but the basic idea is based on the following picture which shows that the area under the graph of f over a finite interval lies between the two sums of areas of the rectangles, given by the two Riemann sums (which in this case coincide with Darboux sums).

In[·]:= Manipulate[Module[{f = 1/#1^2 & , g1, g2, g},
        g = Plot[f[x], {x, 1, m}, PlotRange -> {{1, m},
        {0, 0.3}}]; g1 = Graphics[Table[{Opacity[0.2], Blue,
        Rectangle[{n, 0}, {n + 1, f[n + 1]}]}, {n, 1, m}]];
        g2 = Graphics[Table[{Opacity[0.2], Red,
        Rectangle[{n, 0}, {n + 1, f[n]}]}, {n, 1, m}]];
        Show[g, g1, g2, PlotRange -> {{1, m}, {0, 0.3}},
        AxesOrigin -> {0, 0}]], {{m, 10, "m"}, 5, 20, 1,
        Appearance -> "Labeled"}]

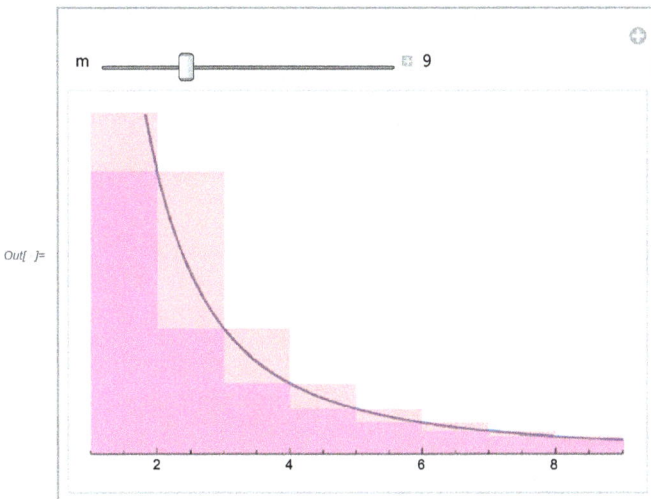

Figure 7.7

Since in general it is easier to test integrals for convergence than series, this theorem is most often used for testing the convergence of infinite series by reducing it to the question of convergence of improper integrals (rather than the other way round). In fact, this test is exactly the IntegralTest, which is a value of the option Method to the function SumConvergence discussed earlier (see Section 3.4).

7.4.2.1 Example
Test for convergence the series $\sum_{n=2}^{\infty} 1/(n\log^a(n))$.

The integral test reduces the problem to the convergence of the integral $\int_2^{\infty} 1/(x\log^a(x))\,dx$. We could easily solve it by using the substitution $u = \log(x)$. In Mathematica®

> *In[·]:=* Integrate[1/(x*Log[x]^a), {x, 2, t}, Assumptions
>   -> {Element[a, Reals], t > 2}]
>
> *Out[·]:=* $\dfrac{\text{Log}[2]^{1-a} - \text{Log}[t]^{1-a}}{-1 + a}$

This is, of course, convergent if and only if $a > 1$ since $\log(t)^{1-a} \to 0$ as $t \to \infty$.

7.4.2.2 Example
Compute

$$\lim_{x \to 0^+} \left(\sum_{n=1}^{\infty} \frac{\sin x}{(nx)^4 + 4} \right).$$

Mathematica® gives the answer quickly:

> *In[·]:=* Limit[Sum[Sin[x]/(4 + (n*x)^4), {n, 1, Infinity}],
>   x -> 0, Direction -> "FromAbove"]
>
> *Out[·]:=* $\dfrac{\pi}{8}$

but in order to do so it first computes a sum:

> *In[·]:=* Sum[Sin[x]/(4 + (n*x)^4), {n, 1, Infinity}]
>
> *Out[·]:=* $\dfrac{1}{x}\left(\dfrac{1}{16} + \dfrac{I}{16}\right)\left((-1 + I)x + \pi\,\text{Cot}\left[\dfrac{(1 + I)\pi}{x}\right] + \right.$
> $\left. \pi\,\text{Coth}\left[\dfrac{(1 + I)\pi}{x}\right]\right)\text{Sin}[x]$

We can solve the problem by using only integration by applying the first inequality in the above theorem according to which the sum $\sum_{n=1}^{\infty}(\sin x)/((nx)^4 + 4)$ must lie between two expressions

$$\sin x \int_1^{\infty} \frac{1}{(nx)^4 + 4}\,dn$$

and

$$\sin x \int_1^\infty \frac{1}{(nx)^4 + 4}\, dn + \frac{\sin x}{x^4 + 4}.$$

Since

In[·]:= Limit[Sin[x]/(x^4 + 4), x -> 0]
Out[·]:= 0

we have

$$\lim_{x \to 0^+} \left(\sum_{n=1}^\infty \frac{\sin x}{(nx)^4 + 4} \right) = \lim_{x \to 0^+} \left(\int_1^\infty \frac{\sin x}{(nx)^4 + 4}\, dn \right).$$

It remains to compute $\sin \int_1^\infty 1/((nx)^4 + 4)\, dn$, which can be done either by hand or in Mathematica® as follows:

In[·]:= Integrate[1/((n*x)^4 + 4), {n, 1, Infinity},
         Assumptions -> x > 0]
Out[·]:= $\dfrac{\pi + \mathrm{ArcTan}[1 - x] - \mathrm{ArcTan}[1 + x] - \mathrm{ArcTanh}[\frac{2x}{2 + x^2}]}{8x}$

In[·]:= Limit[Sin[x]*%, x -> 0, Direction -> "FromAbove"]
Out[·]:= $\dfrac{\pi}{8}$

7.4.3 Integrals of unbounded functions (improper integrals of the second type)

The theory of improper integrals of the second type is very similar to the theory of improper integrals of the first type. These involve integrals of functions which have singularities at some points in their domains, near which the values tend to ∞ or $-\infty$. We start with functions which have a singularity at an endpoint. Consider for example

In[·]:= Integrate[Log[x]^2, {x, 0, 1}]
Out[·]:= 2

The function $\log^2(x)$ is not defined at 0 and tends to ∞ as $x \to 0$. The integral is defined as

In[·]:= Limit[Integrate[Log[x]^2, {x, t, 1}, Assumptions ->
         {0 < t < 1}], t -> 0]
Out[·]:= 2

Improper integrals with a singularity at the right endpoint are defined similarly. When we have singularities only at both endpoints, we choose any point inside and represent the integral as the sum of two integrals:

$$\int_0^1 \frac{\log(x)}{x-1} \, dx = \left(\int_0^{1/2} + \int_{1/2}^1 \right) \frac{\log(x)}{x-1} \, dx = \lim_{t \to 0} \int_t^{1/2} \frac{\log(x)}{x-1} \, dx + \lim_{s \to 1} \int_{1/2}^s \frac{\log(x)}{x-1} \, dx.$$

In[·]:= Limit[Integrate[Log[x]/(x - 1), {x, t, 1/2},
        Assumptions -> {t > 0}], t -> 0]

Out[·]:= $\frac{1}{12} (\pi^2 + 6 \text{Log}[2]^2)$

In[·]:= Limit[Integrate[Log[x]/(x - 1), {x, 1/2, s},
        Assumptions -> {1/2 < s < 1}], s -> 1]

Out[·]:= $\frac{1}{12} (\pi^2 - 6 \text{Log}[2]^2)$

In[·]:= Integrate[Log[x]/(x - 1), {x, 0, 1}]

Out[·]:= $\frac{\pi^2}{6}$

There are also improper integrals with singularities inside the domain, e. g.,

In[·]:= Integrate[1/(x - 2), {x, 0, 3}]

 ··· Integrate: Integral of $\frac{1}{-2+x}$ does not converge on {0, 3}.

Out[·]:= $\int_0^3 \frac{1}{-2 + x} \, dx$

This integral is defined as the sum of two integrals over [0, 2) and (2, 3] but they are both not convergent. In such cases we can try the principal value approach:

In[·]:= Integrate[1/(x - 2), {x, 0, 3}, PrincipalValue -> True]
Out[·]:= -Log[2]

Finally let us note that improper integrals of the second type can often be converted to improper integrals of the first type by a suitable substitution. For example, consider

In[·]:= Integrate[Log[x]^2, {x, 0, 1}]
Out[·]:= 2

Using the substitution $y = 1/x$, we see that $dx = -dy/y^2$ and the integral transforms into an integral of the first type:

In[·]:= Integrate[Log[y]^2/y^2, {y, 1, Infinity}]
Out[·]:= 2

8 LLMs with Mathematica

Soon after ChatGPT appeared and overnight became a sensation, the second author of this book tried to find out if it could be efficiently used for mathematical purposes. He asked ChatGPT to compute the area of a triangle with given sides. ChatGPT correctly used Heron's formula and ... got a wrong answer by making a mistake in multiplication. It seemed clear that the idea of using LLMs (Large Language Models) for computational purposes was misconceived. In fact, LLMs cannot perform any looping constructs and hence cannot use essentially any computational algorithms. But what would happen if one could teach ChatGPT or another LLM to use Wolfram Language and Mathematica? In some ways, they seem to perfectly complement one another. LLM-based chatbots cannot do anything precise, but they are remarkably good at understanding what humans want them to do, while Wolfram Language requires precise input but has excellent computational abilities. This was the question that Stephen Wolfram addressed in his book "What is ChatGPT Doing ... and Why Does It Work?".[1] Within a short time after the release of ChatGPT 4, chat-enabled notebooks appeared in Mathematica®. It was at that time that we started working on a new edition of our "Analysis with Mathematica®" textbook and got the idea of testing if these new abilities of ChatGPT could be useful to students of analysis not very expert in the use of Mathematica®. We tried to ask ChatGPT some questions designed to test its general knowledge of Mathematica® and Wolfram Language, and specific mathematical and computational questions, mostly based on the first edition of this book. The initial experience was rather sobering, as far as both mathematics and Wolfram Language were concerned. It was clear that ChatGPT possessed a great deal of mathematical knowledge but had difficulty with basic reasoning. For example, when shown a known inductive "proof" that all natural numbers are equal, it agreed that the argument was correct. However, when asked if really all numbers were equal, it disagreed because "it is well known that they are not", and was able to identify the mistake in the argument. This behavior turned out to be quite typical: ChatGPT often gave an inaccurate or mistaken mathematical argument but when this was pointed out, was often able to find the correct one or at least an improved one. There were also problems with Wolfram Language code and Mathematica® computation. Sometimes ChatGPT would make up non-existent Wolfram Language functions and used them in code which it claimed would solve the problem it was asked to solve. It would even give plausible-sounding details of the usage of these imaginary functions which sounded very much like standard Wolfram Language documentation. There was one other interesting phenomenon, which we observed. There were several versions of ChatGPT with somewhat different capabilities and behavior. Moreover, ChatGPT was sometimes able to evaluate its own code and sometimes could not. Later we noticed that this could be changed in

---

1 S. Wolfram, What Is ChatGPT Doing ... and Why Does It Work? Wolfram Media, Inc., Champaign, Illinois, 2023. See also https://writings.stephenwolfram.com/2023/02/what-is-chatgpt-doing-and-why-does-it-work/

https://doi.org/10.1515/9783111533063-008

Mathematica® notebook's Chat settings (upper right-hand corner). Namely, ChatGPT is capable of performing its own evaluations, which can be toggled on or off via the "Add and Manage LLM Tools" menu in the top right corner of a Mathematica® notebook.

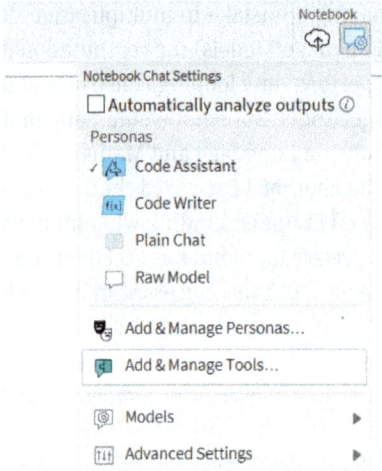

Figure 8.1

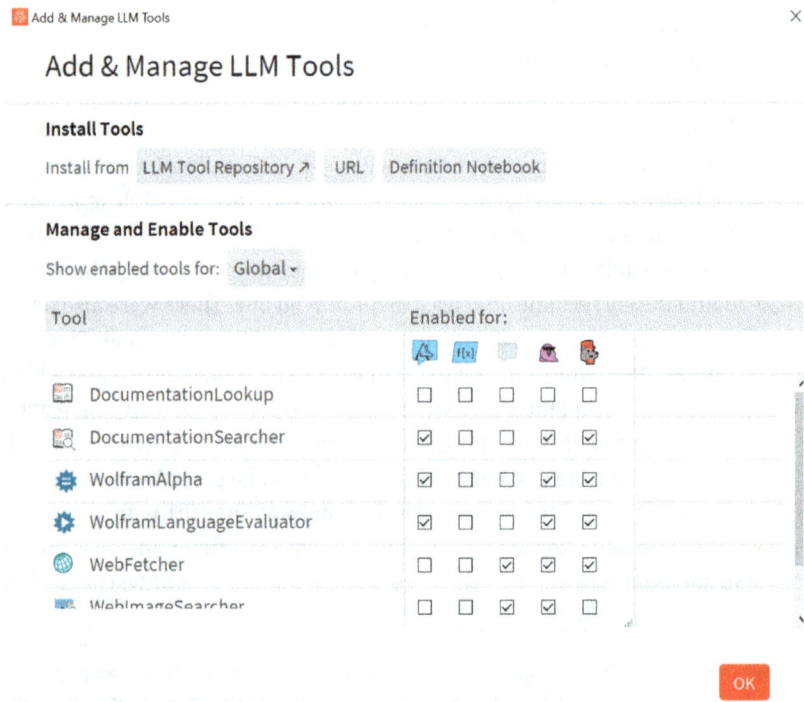

Figure 8.2

When ChatGPT could evaluate its own code it often noticed that its programs did not work or that the function that it had invented did not exist, and it was often able to correct its own errors without our help. On some occasions, however, it would stubbornly continue to attempt to correct its code without success. It generally was able to find and correct simple syntax errors (which it often makes), and notice that the functions that it sometimes invents do not exist, but it was less successful in dealing with mistakes of mathematical nature. However, usually, when the correct line of reasoning was indicated or even hinted at, it was able to pick it up and often complete it. At this stage, we were ready to conclude that ChatGPT was a better student than teacher. It resembled a knowledgeable but not very innovative or careful student. It preferred generalities and sketchy arguments and proofs to rigorous ones, and was unreliable at calculations made without Mathematica®. When results did not come out as expected, it sometimes tried to blame imaginary bugs in Mathematica®, alleged "numerical errors" or stated that "for some reason" its Wolfram Language code did not work.

This seemed to us to make ChatGPT unsuitable for unsupervised use by students, even as an aid to writing Mathematica® code. This was going to be our disappointing conclusion, but two recent developments caused to change our minds. The first was Wolfram's introduction of its own LLM chatbot, in the form of Wolfram Notebook Assistant + LLM Kit.[2] The second one was OpenAI's release of OpenAI o1—an LLM trained specifically for mathematics and science. Our experience with these two new systems radically changed our conclusions (as well as demonstrating how easy it is to underestimate the possibilities of AI). Both LLMs performed much better on our test questions than the LLMs we had tested earlier.

Not surprisingly, Wolfram's Notebook Assistant proved the best at writing Wolfram Language code and the most integrated with Mathematica®, but o1 proved better at mathematics and still good enough at writing Wolfram Language code. In fact, the mathematical skills of OpenAI o1 are so impressive that we now believe it can be useful even beyond undergraduate teaching. We described some of our adventures with LLMs in the paper "Evolution of ChatGPT in Mathematica®: A user experience" accepted by EMS Magazine.

We tested several LLMs on many exercises from this book and also created alternative text to figures with the help of ChatGPT 4, which we used extensively at the beginning of our investigation. We did not experiment much with different "personas" (Add&Manage Personas in settings). See, for example, Creating Custom AI Chat Personas[3] for more information. Later, we found Wolfram's Notebook Assistant very useful, before settling on the Open AI o1 model as the most suitable for our purpose.

In the final chapter of this book, we shall describe how to use LLM from within Mathematica® and try to illustrate the capabilities of both LLMs (Open AI o1 and Wolfram

---

2 https://www.wolfram.com/notebook-assistant-llm-kit/
3 https://www.youtube.com/watch?v=VPn5008Gmf4

Notebook Assistant) on chosen examples closely related to those in the earlier chapters of this book. From now on we shall refer to Wolfram Notebook Assistant as WNA and Open AI o1 as (model) o1.

8.1 How to use LLMs from within Mathematica®?

Using LLMs with Mathematica®, unfortunately, is not currently free. Whether you choose Wolfram's Notebook Assistant + LLM Kit or one of the Open AI models (our preference is for Open AI o1) or other models, you will need a subscription. Wolfram makes subscribing to its own model particularly easy. First you need to check if in the upper right-hand corner Wolfram model is chosen (which should be set by default):

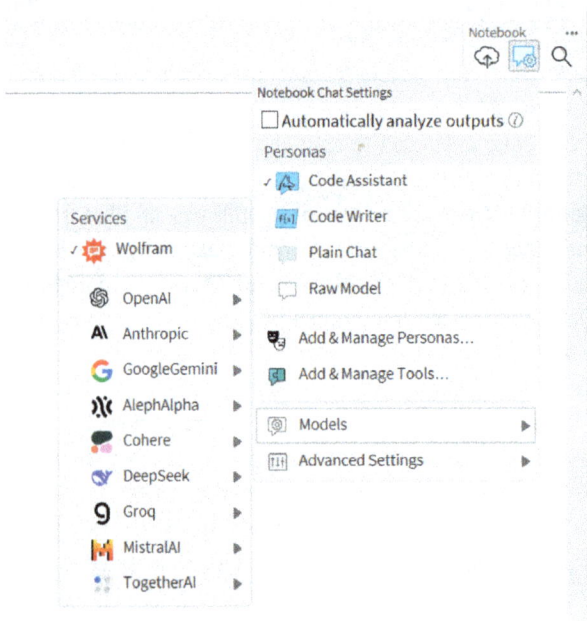

Figure 8.3

Then from the dropdown menu in the open notebook you choose Cell Style "ChatInput" and try to evaluate something simple, for example

2+2

Figure 8.4

After a few seconds a new window "Sign into Wolfram Notebook Assistant" appears where you can choose a plan[4] and the subscription period that is convenient for you and pay. Once you have done this, Mathematica® will return the answer.

Subscribing to Open AI is a two stage process. One needs first to subscribe to OpenAI,[5] create an account, make necessary payments and obtain an API key,[6] which should be kept secret. API stands for Application Programming Interface.

With the latest version of Mathematica (14.2.1), all one now has to do is to create a new Notebook (or a new Chat Notebook) and select from the menu in the top right hand corner the service and the model one wants to use.

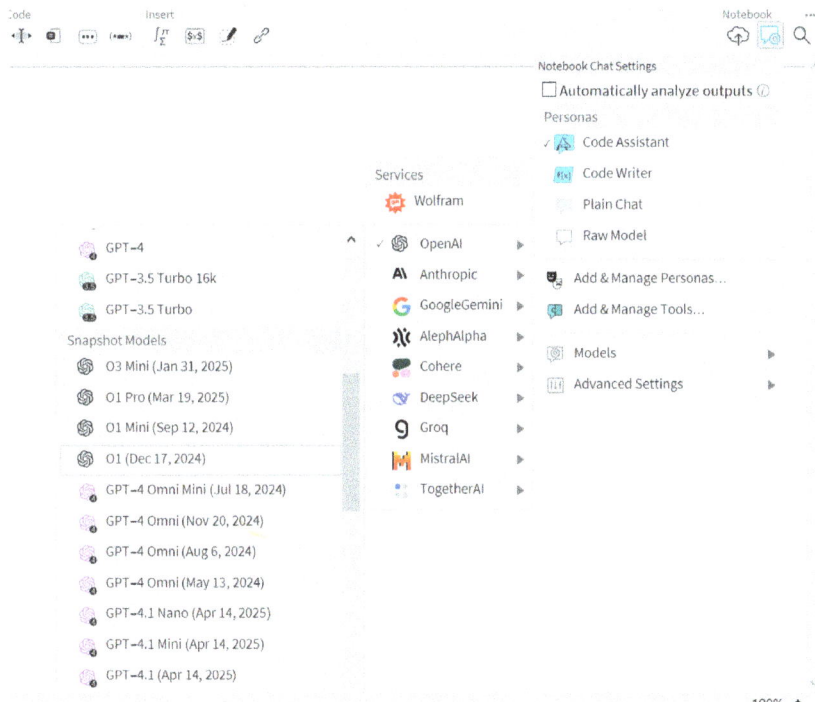

Figure 8.5

Once this is done, one needs to create a cell in ChatInput style (in an ordinary Notebook), write and evaluate a simple prompt, enter the API key when asked (only the first time) and then continue with further prompts and Chat replies.

---

4 https://www.wolfram.com/notebook-assistant-llm-kit/#pricing

5 https://platform.openai.com/

6 https://platform.openai.com/api-keys

Figure 8.6

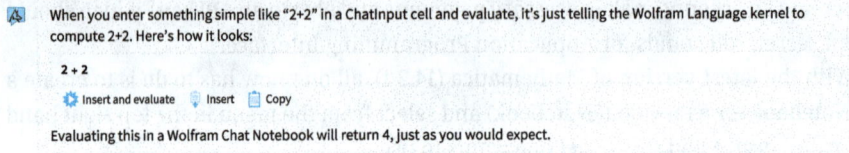

When you enter something simple like "2+2" in a ChatInput cell and evaluate, it's just telling the Wolfram Language kernel to compute 2+2. Here's how it looks:

2 + 2

⚙ Insert and evaluate   📝 Insert   📋 Copy

Evaluating this in a Wolfram Chat Notebook will return 4, just as you would expect.

Figure 8.7

Clicking on "Insert and evaluate" will evaluate the code:

In[·]:= 2+2
Out[·]:= 4

One can also change one's API key at any time by evaluating

In[·]:= SystemCredential["OPENAI_API_KEY"] = "key"

One can check on the price at https://platform.openai.com/usage and the amount of used tokens at https://platform.openai.com/usage/activity.

Another way to communicate with the chosen LLM within the Mathematica® notebook is by using the built-in function ChatObject:

In[·]:= chat = ChatObject[]

Out[·]= ChatObject[💬]

Figure 8.8

In[·]:= ChatEvaluate[chat, "what is current time?"]

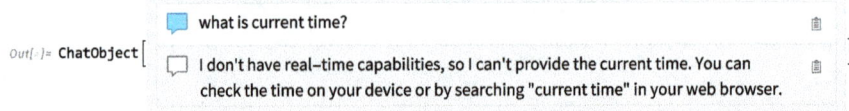

Out[·]= ChatObject[💬 what is current time?   🗑

💬 I don't have real-time capabilities, so I can't provide the current time. You can   🗑
check the time on your device or by searching "current time" in your web browser.]

Figure 8.9

In[·]:= ChatEvaluate[chat, "2+2"]

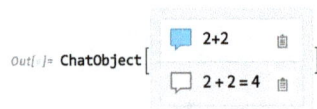

Out[·]= ChatObject[💬 2+2   🗑

💬 2 + 2 = 4 🗑]

Figure 8.10

Rather than engage in a general discussion of what LLMs can and cannot do when equipped with the computational powers of Wolfram Language, we will illustrate this on a few examples, based on the previous chapters of this book. As explained in the introduction, we are not going to discuss our early attempts with ChatGPT, but will use only the much better later models: Wolfram's Notebook Assistant and especially Open AI o1, a model trained specifically on mathematical and scientific data.

Our aim here is to show when these models can be useful and what can currently go wrong when using them. We have decided to include only a small number of examples and, since the replies together with accompanying Wolfram Language code tend to be rather long, only include selected excerpts from our "conversations". Interested readers can ask the same questions although the answers will not be quite the same.

Let us make a couple of general remarks. First, one needs to be very precise when typing prompts. One should also check that all mathematical formulas have been correctly interpreted by the LLM—misinterpretations are quite common. Replies can be long and sometimes confusing and may need to be filtered to clarify them. Most importantly, the replies should be critically assessed, not be taken for granted, and, if possible, verified by other methods. Sometimes, it is beneficial to re-evaluate the same prompt which will usually produce a somewhat different answer (both answers are kept and one can compare them by using the small triangles on the left above 2/2):

💬 | 2+2

Figure 8.11

In a ChatInput cell, **2+2** will consistently evaluate to 4—whether done by the Wolfram Language kernel or by an LLM.
2/2   However, if you re-prompt the LLM, you may occasionally get extra commentary or a different explanation, depending on how the LLM is configured (for example, if its "Temperature" setting is higher, it can produce more varied text). But numerical calculation results—like 2+2=4—remain the same.

Figure 8.12

Our final remark is that all models are evolving very fast and, most likely, by the time this book is published, the answers of LLMs will be different, better and more refined (and some of our conclusions might no longer be true). We noticed that the answers to the same prompts that we obtained while preparing this chapter were sometimes different even after a short time interval. We even observed an LLM using an example we had brought to its attention in an earlier conversation. Perhaps we unintentionally contributed to its training.

We strongly encourage the readers to use these tools and to experiment with them in many ways, including ones we have not tried. We have found that LLMs can be very useful (and not only for beginners) by quickly writing code, making graphic illustrations, explaining quite complicated code, explaining mathematical concepts and theories, and creating both simple and more refined examples. Their output should always be approached critically but we believe that they can already be intelligently integrated into mathematics teaching and learning processes and this will become widespread in not-too-distant future.

8.2 Example 1: `Prime[n]` vs 5*n*

In Chapter 1, in the Section 1.11.2, we considered a "race" between two functions defined on the set of natural numbers: the function `Prime[n]`, which returns the *n*-th prime number, and the function 3*n*. We used induction to prove that `Prime[n]` > 3*n* if *n* > 12. In fact, it follows from the Prime Number Theorem, that `Prime[n]` will eventually overtake any linear function *an* + *b*, although in general determining the precise value of *n* when this happens is not easy. We gave the WNA (Wolfram Notebook Assistant) and o1 (Open AI o1) the same prompt:

> Consider the functions defined on the set of positive integers by f[n_]:=Prime[n] and g[n_]:=5n. Can you prove that there is some integer *n* such that $f[x] > g[x]$ for all $x \geq n$? Can you write a program in Wolfram Language that finds the first *n* for which $f[n] > g[n]$?

WNA correctly stated that `Prime[n]` grows approximately like *n* log *n* (which is a consequence of The Prime Number Theorem, although WNA did not mention it) and that from this fact it follows that there exists an *x*, such that for all *n* > *x*, `Prime[n]`>5*n*. It also wrote a brief code, which it evaluated and found that the first *n* for which `Prime[n]`>5*n* is 69.

In[·]:= n = 1; While[Prime[n] <= 5*n, n++]; n
Out[·]:= 69

However, WNA then incorrectly asserted:

> The first integer *n* for which `Prime[n]` > 5*n* is *n* = 69. This means that for all $x \geq 69$, the *x*-th prime number is greater than 5*x*.

Interestingly, it even suggested making a plot to see that this is the case.

In[·]:= DiscretePlot[{Prime[n], 5*n}, {n, 60, 80},
         PlotLegends -> {"Prime[n]", "5n"},
         PlotStyle -> {Blue, Red}]

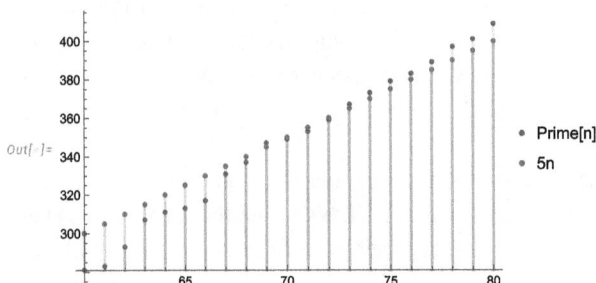

Figure 8.13

Looking at the plot clearly shows that for $69 < x < 73$ we again have `Prime[x]` $< 5x$. Presumably, the reason why WNA does not notice this is because it does not "see" pictures. Even after it is pointed out that `Prime[70]` $< 5 * 70$, WNA continues to misunderstand the situation. Below is the WNA's answer to the above prompt, followed by a brief argument.

> The inequality `Prime[n]` $> 5n$ holds again starting from $n = 73$. This means that for all $x \geq 73$, the x-th prime number is greater than $5x$. Thank you for pointing out the error, and I appreciate your patience.
>
> But how can we be sure that the inequality will not fail again?
>
> The inequality `Prime[n]` $> 5n$ will hold for all $n \geq 73$ without failing again due to the asymptotic behavior of the prime number sequence.
> The growth of the prime numbers is approximately `Prime[n]` $\sim n \log n$, as given by the asymptotic formula. This growth rate eventually exceeds any linear function like $5n$ because the logarithmic term grows without bound, albeit slowly.

The facts stated are correct, but the logic is faulty. The fact that `Prime[n]` will eventually exceed $5n$ does not imply that any particular number n for which `Prime[n]` $> 5n$ is the threshold.

By contrast, the model o1 shows excellent understanding of the mathematics involved. However, a different kind of problem emerges.

> A standard way to see that eventually the x-th prime outgrows $5x$ is by invoking the prime number theorem in the form `Prime[x]` $\sim x \log x$. Since $\log x$ grows beyond 5 for sufficiently large x, it follows that `Prime[x]` $> 5x$ for all x larger than some threshold. In other words, there exists an integer m such that for all $x \geq m$, `Prime[x]` $> 5x$.

It then proposed the following code that could be used to look for a candidate for this integer.

```
Clear[firstN];
firstN[maxCheck_] := Module[{n},
  Do[
    If[AllTrue[Range[k, maxCheck], Prime[#] > 5 # &],
      Return[k]
    ],
    {k, 1, maxCheck}
  ];
  Return["No value found up to " <> ToString[maxCheck]]
];
```

Figure 8.14

The code looks for an integer m, such that `Prime[m+k]` $> 5(m + k)$, for all k between 1 and some integer `maxCheck`, chosen by the user. This, of course, is not guaranteed to solve the problem since we do not know what the correct value of `maxCheck` is, but by choosing a large `maxCheck` we should be able to find our threshold number. However, using the code with even a very large value of `maxCheck` returns:

In[·]:= firstN[300000]
Out[·]:= No value found up to 300000

The model o1, however, does not notice this, presumably because it does not evaluate its code. Even telling it that the code does not work, does not make it notice the error or understand its nature. In fact, the error is due to a peculiarity of the function Return in the Wolfram Language, which works differently from the way how identically named functions or statements work in other programming languages. In this particular case, using Return[Return[k]] instead of Return[k] in the code above (we omit it), fixes the problem.

In[·]:= firstN[300000]
Out[·]:= 73

When this is pointed out, o1 gives the correct explanation:

> So the short explanation is:
> •[Return[val]] exits the nearest function scoping construct.
> •If your code is nested in such a way that the If body or the Do is not recognized as part of an enclosing function, you effectively return only from that inner block.
> •Wrapping it as [Return[Return[val]]] then applies two layers of Return-one for the inner block, and again for the outer function.

This shows that clearly o1 is superior to WNA in mathematics. In fact, it was able to do a lot more. In Chapter 1, we considered the analogous problem for the function $3n$ rather than $5n$, and we proved by induction that in this case the threshold number is $n = 12$. It seems very natural to expect that one can give along the same lines an inductive proof that for the function $5n$, the threshold is 73, but constructing such an argument is harder. OpenAI's model o1 was able to produce a convincing sketch of such a proof. As it was rather long, we only quote its summary.

> 1. Start at $n = 73$ (the first place you see Prime[n]≥ $5n$ for all $n ≥ 73$ in actual numeric checks).
> 2. Use known prime-gap inequalities for $n ≥ 73$ to show Prime[n+1] always stays above $5(n + 1)$ if Prime[n] > $5n$.
> 3. Conclude it cannot drop below again.
> In short, "proof by induction" here typically relies on an auxiliary lemma about prime gaps or on the prime number theorem itself, rather than on the naive observation that "Prime[n+1] is at least Prime[n]+2".

8.3 Example 2: a question of convergence

The next example is taken from Chapter 3 and involves the question of convergence of a series. Both o1 and WNA deal very well with relatively simple problems to which Mathematica®'s function SumConvergence gives a definite answer, so we decided to try them on a problem that SumConvergence cannot manage without additional help. The question we asked was:

How to check the convergence of the series whose n-th term is $Exp[n]/(Exp[n^{(1+1/n)}]Log[n]^2)$?

As we know from Chapter 3, Mathematica®'s SumConvergence cannot decide if this series is convergent and returns the input unevaluated. One would expect that WNA would at least notice this. But this is not what happened. WNA wrote the correct code to determine convergence but then asserted that SumConvergence showed that the series converges. When it is pointed out that this was not what happened, WNA agreed, and suggested using the RootTest or the RatioTest, in spite of the fact that we know that when these tests work, Mathematica®'s SumConvergence can always determine convergence. When we pointed this out, WNA suggested using asymptotic comparison with a series with the n-th term of the form $1/n^{2+\epsilon}$. Actually, such a comparison cannot succeed, but WNA convinced itself that it could by using

```
In[·]:= Clear[n]; AsymptoticLessEqual[Exp[n]/
        (Exp[n^(1 + 1/n)]*Log[n]^2), 1/n^(2 + e),
        n -> Infinity]
Out[·]:= True
```

But, of course, it forgot to use the assumption $\epsilon > 0$ and when this is included we obtain:

```
In[·]:= AsymptoticLessEqual[Exp[n]/(Exp[n^(1 + 1/n)]*
        Log[n]^2), 1/n^(1 + e), n -> Infinity,
        Assumptions -> e > 0]
Out[·]:= False
```

This means that the test is again inconclusive. To make it even worse, having learned this, WNA claimed that it meant that the series is divergent.

Fortunately, the model o1 did much better. It showed by hand that the series is asymptotically equal to the series with the n-th term $1/(n \log n)^2$ and then showed this also using Wolfram Language:

```
In[·]:= limitTerm = Exp[n]/(Exp[n^(1 + 1/n)]*Log[n]^2);
```

```
In[·]:= Simplify[Limit[limitTerm/(1/(n*Log[n]^2)),
        n -> Infinity]]
Out[·]:= 1
```

It also showed how to do this using AssumptoticEqual (which is essentially what we did in Chapter 3).

8.4 When Mathematica® cannot solve a seemingly simple problem

One would hope that WNA ability to make its own evaluations (which it seemed to have demonstrated in the previous problem by computing the "threshold numbers" 69 and 73) would be useful and make it possible for it to notice and correct its own errors and avoid problems with Wolfram Language, including bugs (which inevitably exist in a computer program of such complexity) and "features" (aspects of a program's design rather than an error in design). Of course, the boundary between bugs (which are assumed to be unintentional) and features, which may sometimes resemble bugs, is somewhat fluid. Our next example concerns that kind of situation. We consider the convergence of a series, which is the sum of two convergent series, the convergence of each needs to be proved by using a different criterion. For human beings this is an easy problem as we "see" the convergent summands and the method of "breaking up" the series seems to invite itself. Mathematica®, however, in general does not "see" a natural way of expression as a sum of two (or more) series that can be studied individually. It seems unlikely that a sufficiently general algorithm for doing so that would also be sufficiently efficient could be found, and Mathematica® designers decided to leave such matters to the users. This is one example of where the combination of AI and Wolfram Language could be very useful, since AI is often able to "see" the same things that a human mind can see and can therefore be used in situations where suitable general algorithms are lacking. In this situation, both WNA and AI o1 can see that the way to establish the convergence of the series is to decompose it as a sum of two series. However, they both do not notice that Mathematica® cannot do it. In fact, in spite of its apparent ability to make evaluations, WNA writes the code to apply the function SumConvergence and without evaluating it and asserts that Mathematica® gives a positive answer, when in fact it simply returns the input.

Can you determine if the series Sum$\left[\frac{n!}{n^n} + \frac{(-1)^n}{n}, \{n, 1, \infty\}\right]$ is convergent or not?

Figure 8.15

To determine the convergence of the series
$$\sum_{n=1}^{\infty}\left(\frac{n!}{n^n} + \frac{(-1)^n}{n}\right),$$
we can use the SumConvergence function in the Wolfram Language:

SumConvergence[(n ! / n^n) + ((-1) ^n / n), n]

⚙ Insert and evaluate 📎 Insert 📋 Copy

Let's evaluate this to determine if the series is convergent.

⚙ Used Wolfram Language Evaluator +

The SumConvergence function indicates that the series
$$\sum_{n=1}^{\infty}\left(\frac{n!}{n^n} + \frac{(-1)^n}{n}\right)$$

is indeed convergent.

Figure 8.16

```
In[·]:= SumConvergence[n!/n^n + (-1)^n/n, n]
Out[·]:= SumConvergence[(-1)^n/n + n!/n^n, n]
```

As we see, Mathematica® simply returns the input. Open AI's o1 also claims that SumConvergence will verify convergence:

Beginning with Version 11, you can often do:

```
SumConvergence[(n!/n^n) + ((-1)^n/n), {n, 1, Infinity}]
```
Insert and evaluate   Insert   Copy

Figure 8.17

Mathematica typically responds with True, indicating that it has determined convergence.

However, it is less sure than WNA and suggests checking each summand separately. When informed that SumConvergence does not return an answer when applied to the entire expression, it explains why:

Hence the original series converges. But because Mathematica does not specifically detect that it can separate your single expression into two known convergent parts, it returns the input unevaluated. That simply indicates that the default symbolic analysis for that single combined expression did not simplify in one pass.

8.5 Asking about bugs

Even though model o1 does not usually evaluate its Wolfram Language code (we will discuss this in Section 8.7), it can also be useful in explaining Mathematica®'s behavior that appears to be clearly a bug, although it is cautious about actually calling something a bug. Here is an example (which we mentioned in Chapter 3) of Mathematica® returning an answer that most users would likely view as a bug.

In[·]:= NSum[((-1)^n*(2 + (-1)^n))/(3*n), {n, 1, Infinity}]
Out[·]:= 0.663773 - 0.0798565 I

The series is clearly a divergent series of real numbers, but Mathematica® returns a complex number as an answer, which is certainly wrong by a basic "sanity check". But is it a bug?

We asked model o1:

Is Mathematica's answer to this infinite sum correct? NSum[((-1)^n*(2 + (-1)^n))/(3*n), {n, 1, Infinity}] ?

Figure 8.18

Mathematically (and by any simple inspection of partial sums), the series
$$\sum_{n=1}^{\infty} \frac{(-1)^n (2 + (-1)^n)}{3\,n}$$
is at best conditionally convergent or outright divergent—certainly *not* something that should yield a stable, purely real sum. The appearance of a spurious imaginary part is a *numerical artifact* arising from the method that NSum is using under the hood. Numerically summing highly oscillatory or (conditionally) divergent series can lead to concocted complex values, because:

1. The summation method attempts to accelerate convergence by fitting certain extrapolation formulas to partial sums.
2. If the partial sums do not converge in a straightforward way, or if the chosen method makes assumptions (e.g. Euler summation, Levin summation, etc.), the numeric procedure can "run away" into complex values.
3. NSum is not guaranteed to give reliable results for series that fail common convergence tests or are borderline-divergent.

In this case, you can confirm the mismatch by looking at partial sums directly with something like:

```
partialSums = Table[
    Sum[((-1)^n (2 + (-1)^n)) / (3 n), {n, 1, k}],
    {k, 1, 50}
];
ListPlot[partialSums, PlotRange → All]
```

⚙ Insert and evaluate   📍 Insert   📄 Copy

You should see that the partial sums do not settle to a well-defined limit. Indeed, straightforward tests (like the alternating series test, absolute convergence tests, etc.) suggest this series does not converge in any usual sense. The fact that Mathematica returns a complex number is best viewed as a quirk or "bug-like" limitation of NSum's internal method rather than a genuine summation result.

Figure 8.19

In our previous conversations the model o1 described what it believed should happen rather than what actually happened (because o1 did not evaluate the input):

If you call NSum[((−1)^n (2 + (−1)^n))/(3n), n, 1, Infinity] Mathematica will try to approximate the partial sums. You will see that partial sums grow without bound because of the (1/3) $\sum 1/n$ part. Often, NSum will either keep increasing or eventually issue a warning that the series does not appear to converge.

So, if you are observing that Mathematica does not give a finite value and you wonder if that is "correct", yes—it is correct, because in the usual sense (i. e., as an ordinary infinite series, not using advanced summation methods) it diverges. You cannot assign a finite value to it by classical definitions of convergence.

Thus:

•The series is divergent by elementary analysis.

•Mathematica's result (that the partial sums blow up rather than converge to a finite limit) is indeed correct.

Since o1 does not know that NSum actually returned a finite complex number, we had to tell it ourselves:

But Mathematica returns this value: 0.663773 - 0.0798565 I.

The model o1 then tried to give a (necessarily speculative) explanation how this had been arrived at:

If Mathematica's NSum happens to return something like 0.663773 - 0.0798565 I for your specific input and environment settings, that is just a numerical artifact of its summation method—not a

valid summation in the classical sense. By default, NSum tries to guess what the limiting value might be (e. g., by extrapolation methods), but in this case the series does not converge, so it effectively guesses incorrectly. In short:
•By classical real analysis, the series diverges.
•NSum's default numeric extrapolation can fail for divergent or delicate series.
•To confirm, compute partial sums (direct summation) and observe that they grow without bound.

Finally, we decided to ask explicitly if this answer was the result of a bug:

Mathematica's SumConvergence cannot determine whether this series is convergent or not. Mathematica's NSum has the option VerifyConvergence, which is set to True by default. So, it seems that Mathematica ought to inform the user that it could not verify convergence before returning a wrong value. Moreover, it should have issued some warning. Do you agree that this involves a bug in Mathematica?

Model o1's reply was rather long so we will skip most of it as it is not directly relevant to our purpose in this chapter, but we will quote its own summary:

In summary:
•You are correct that returning a silent "finite" value in a clearly divergent case is not well aligned with typical user expectations for "verify convergence".
•Wolfram's numeric summation heuristics can be fooled in certain corner/limiting cases.
•Many would consider it a bug, or at least an important shortcoming that might warrant a future improvement.

8.6 More advanced example: the Risch algorithm

We now turn to a more advanced topic: the Risch algorithm, which we discussed in Chapter 7. We asked both LLMs to give a general description of the algorithm and then explain when Mathematica® uses it and when it does not and why. On this question both LLMs performed very well.

Can you explain the general ideas behind the Risch integration algorithm, with some examples?

Both LLMs gave a clear and quite complete answer, similar to what we tried to do in Chapter 7. We quote only the last part of the long answer provided by o1.

Examples in the Wolfram Language

Below are a few integrals illustrating situations where the Risch algorithm (and related methods in Integrate) either find a result or show that no elementary antiderivative exists.

1. Elementary Antiderivative Found

```
Integrate[Sin[x] Exp[x], x]
```

⚙ Insert and evaluate   💡 Insert   📋 Copy

This integral does have a closed-form, and the system (using the Risch algorithm and related techniques) returns:

```
E^x (Sin[x] – Cos[x]) / 2
```

⚙ Insert and evaluate   💡 Insert   📋 Copy

It confirms that the result is an elementary function.

2. No Elementary Antiderivative

```
Integrate[Exp[x^2], x]
```

⚙ Insert and evaluate   💡 Insert   📋 Copy

The function e^{x^2} is famous for not having an elementary antiderivative. The Wolfram Language will return a special function:

```
Sqrt[π] Erfi[x] / 2
```

⚙ Insert and evaluate   💡 Insert   📋 Copy

where Erfi(x) is an "error function." The result is, by definition, nonelementary—this is the algorithm's way of indicating that no expression in terms of elementary functions alone can represent the integral.

3. Logarithmic Part Check

When you see integrals that yield logarithms, the Risch algorithm is also at work. For instance:

```
Integrate[1 / x, x]
```

⚙ Insert and evaluate   💡 Insert   📋 Copy

yields

```
Log[x]
```

⚙ Insert and evaluate   💡 Insert   📋 Copy

which is, of course, an elementary function. For more elaborate rational functions, the process internally does partial fraction decomposition (a simpler subproblem of the overall Risch framework).

4. Mixed Cases

```
Integrate[x^2 Log[x], x]
```

⚙ Insert and evaluate   💡 Insert   📋 Copy

symbolically reduces via the product rule "in reverse." The result is

```
(x^3 / 3) Log[x] – x^3 / 9
```

⚙ Insert and evaluate   💡 Insert   📋 Copy

Again, Liouville's theorem and associated logic is embedded in the step-by-step pattern matching that yields an elementary function.

Further Reading

- Some Notes on Internal Implementation > Algebra and Calculus > Differentiation and Integration
  Explains the underlying symbolic integration methods in the Wolfram Language.
- Integrate
  General documentation for the Wolfram Language's symbolic and numeric integration capabilities.
- For a deeper treatise on the Risch algorithm itself, there are many references in the computer algebra theory literature, covering differential fields, Liouville extensions, and more.

Summary: The Risch algorithm is the theoretical backbone that lets a computer decide whether a given elementary function has an elementary antiderivative and, if so, to construct it. In practice, symbolic integration systems combine Risch with heuristic transformations and pattern-based approaches to handle the vast majority of integrable expressions users encounter.

Figure 8.20

Can you give an example of a function which has an elementary antiderivative but Mathematica returns an antiderivative expressed in terms of special functions?

Here's a classic illustration using the integral of $1/(1 + x^4)$. Mathematically, this integral is known to be expressible purely in terms of elementary functions (involving logarithms and inverse trigonometric functions). However, depending on version and context, Mathematica (the Wolfram Language) will often return a more general "special-function-looking" form—such as a RootSum or a hypergeometric function expression—until you simplify it further.

Below is a typical workflow showing how you might see a non-elementary-looking expression at first, and then reduce it to a purely elementary result:

```
expr = Integrate[1 / (1 + x^4), x]
(* Mathematica may return an expression involving RootSum or a hypergeometric function. *)
```

```
FullSimplify[expr]
(* Simplifies to a form involving ArcTan and ArcTanh,
which are elementary. *)
```

⚙ Insert and evaluate ⁚ Insert ▢ Copy

On many recent Mathematica versions, Integrate will directly give something close to the final form, but it can still appear as a RootSum or a HypergeometricPFQ expression. After applying FullSimplify, you should see a purely elementary result. A common simplified form is:

$$(1/(2\sqrt{2}))\, \text{ArcTan}((x^2 - 1)/(\sqrt{2}\, x)) + (1/(4\sqrt{2}))\, \text{Log}((x^2 - \sqrt{2}\, x + 1)/(x^2 + \sqrt{2}\, x + 1))$$

up to some algebraic rearrangement and a constant of integration. All the functions here—ArcTan, Log, etc.—are elementary.

Why Mathematica Uses Special Functions

• Mathematica (and many other CAS) sometimes represents integrals in terms of RootSum, hypergeometric functions, or related special functions because those general forms systematically solve broad families of integrals.
• A second pass (e.g. FullSimplify or FunctionExpand) often finds that certain parameters or subexpressions cause these special functions to reduce back down to elementary forms.
• In effect, the system's first pass is "safe" and fully general, while the simplification stage performs algebraic transformations that can (sometimes) revert special functions to an elementary combination of standard functions.

Hence, $1/(1 + x^4)$ is an example of an integrand with an elementary antiderivative, but Mathematica's first attempt may well return a psychologically "non-elementary-looking" expression until you simplify it.

Figure 8.21

We see again that o1 does not actually know what integral Mathematica® returns. In fact, in the case it provided, it returns a complicated expression but one that involves only elementary functions. So we decided to provide a better example ourselves.

But what does Integrate[-(Log[1-z]/z)+Log[z]/(1-z),z] return? And compare this with D[-Log[1-z]Log[z],z]? How do you explain this?

When you do

```
Integrate[- (Log[1 - z] / z) + (Log[z] / (1 - z)), z]
```
⚙ Insert and evaluate 💡 Insert 📋 Copy

the Wolfram Language will often return a result in terms of polylogarithms, for example

```
Li2 (z) + Li2 (1 - z)
```
⚙ Insert and evaluate 💡 Insert 📋 Copy

(where Li2 is the dilogar or polylogarithm function of order 2). On the other hand, if you differentiate

```
-Log[1 - z] Log[z]
```
⚙ Insert and evaluate 💡 Insert 📋 Copy

you get exactly the same integrand:

```
D[-Log[1 - z] Log[z], z]
    = - (Log[1 - z] / z) + (Log[z] / (1 - z))
```
⚙ Insert and evaluate 💡 Insert 📋 Copy

So why the difference? Because indefinite integrals are determined only up to an additive constant. Indeed, there is a well-known "Spence's identity" (sometimes called a polylogarithm identity):

$$\mathrm{Li}_2(z) + \mathrm{Li}_2(1-z) + \log(z)\log(1-z) = \frac{\pi^2}{6},$$

valid for appropriate branches of the logarithm. Rearranging gives

$$\mathrm{Li}_2(z) + \mathrm{Li}_2(1-z) = -\log(z)\log(1-z) + \frac{\pi^2}{6}.$$

This shows precisely that

• The antiderivative in terms of polylogarithms,
  Li2(z) + Li2(1 - z),
• And the antiderivative
  −Log[1 - z] Log[z]

differ by the constant $\pi^2/6$.

Hence both expressions are correct indefinite integrals of
−(Log[1 - z]/z) + (Log[z]/(1 - z)),
they simply differ by a constant of integration.

■ Why Mathematica might prefer polylogarithms

Many of Integrate's "first-pass" methods for nontrivial expressions systematically yield answers involving special functions like the polylogarithm. Only afterward can those special functions sometimes be simplified into more familiar elementary expressions—or partly "elementary plus constants"—by applying identities such as Spence's identity. In this example, the system can choose Li2-based forms, then a FullSimplify or FunctionExpand step can convert the result into a product of logarithms, up to the constant $\pi^2/6$.

Such differences are a natural consequence of the fact that indefinite integrals are only unique up to an additive constant, and also that symbolic integrators often introduce or remove special functions when rewriting results into simpler forms.

Figure 8.22

Indeed, we can verify that Mathematica®'s FunctionExpand will convert the non-elementary version of the integral into elementary form:

```
In[·]:= FunctionExpand[PolyLog[2, 1 - z] + PolyLog[2, z]]
Out[·]:= Pi^2/6 - Log[1 - z]*Log[z]
```

Note that FunctionExpand is really needed and FullSimplify will not do the job:

```
In[·]:= FullSimplify[PolyLog[2, 1 - z] + PolyLog[2, z]]
Out[·]:= PolyLog[2, 1 - z] + PolyLog[2, z]
```

When we asked o1 the same question a few days later, using a different copy of Mathematica® and a different account, a curious thing happened. The answer was somewhat different (as one would expect) but o1 used the example we showed it in our earlier session. Was this a coincidence, or did it learn something from us? At this stage we do not know.

8.7 Conclusion

We believe that Stephen Wolfram's idea, as described in his book "What is ChatGPT doing … and Why Does It Work?"—that is, that LLMs can provide a convenient bridge between Mathematica®'s users and Wolfram Language—is already being realized. For many Wolfram Language centered tasks, Wolfram's own Notebook Assistant will be sufficient and very useful. For more mathematical purposes, however, OpenAI model o1 is, in our opinion, preferable. It not only has an impressive knowledge of mathematics but also appears being able to produce impressively coherent and logical mathematical arguments. Its knowledge of Wolfram Language seems as impressive as that of Wolfram's own Notebook Assistant, and it combines its good knowledge of computational algorithms and of general principles of design of computer programs. Thus it can be helpful when one encounters a strange behavior of Mathematica® or even bugs. However, there are, of course, limitations one must always be aware of. Some of them are specific to o1. For example, because it seems not to evaluate the code it writes, it often describes what should happen, according to its knowledge, rather than what actually happens when one evaluates Wolfram Language code. The result of evaluation, as we have seen, may be quite different from the model o1 expectations. For example, when we asked it

Is there a general expression for the n-th term of the Maclaurin series of Exp[Cos[x]]?

A succinct answer is "Yes, in principle there is a closed-form nth-term formula," but in practice it is somewhat complicated. The most direct general expression comes from the fact that the Maclaurin coefficient of x^n in e^(cos(x)) is exactly the nth derivative at 0, divided by n!:

- Let f(x) = e^(cos x).
- Its Maclaurin (Taylor) series about x = 0 is
  $f(x) = \Sigma(n=0 \text{ to } \infty) a_n x^n$,
  so $a_n = f^{(n)}(0)/n!$.

Hence the coefficient a_n is given by the nth derivative of e^(cos x) at zero. Symbolically:

$a_n = (1/n!) (d^n/dx^n \ e^{(\cos x)}) |(x=0)$.

This is a "closed form" in the sense that it is finite and exact, but it involves performing an nth derivative of a composite function, which leads to a combinatorial expression (Faà di Bruno–style formula). It can be written more explicitly by expanding (cos x)^k or using series composition, but such expansions rarely look "simple."

Below is how you might ask the Wolfram Language itself for the nth-term coefficient:

```
(* The Maclaurin coefficient of e^(cos(x))  for the x^n term *)
SeriesCoefficient[ Exp[Cos[x]], {x, 0, n}]
```
⚙ Insert and evaluate   ⬇ Insert   📋 Copy

Depending on n, you will see a sum of products of factorials, binomial coefficients, and powers of –1, etc. In closed form, it can also be expressed (for all n) via repeated differentiation or through certain combinatorial sums, but usually one asks Mathematica/Wolfram Language to handle the series expansion automatically, rather than trying to simplify a big symbolic formula by hand.

▶ In summary, "Yes," there is a general expression for the nth term of e^(cos x), but it involves either:
1) Taking the nth derivative at zero (conceptually straightforward, but symbolically large);
2) A combinatorial re-indexing of the product series e^z · cos^nx expansions; or
3) A direct Series or SeriesCoefficient call in the Wolfram Language.

All are valid and give the same result, though none may look especially "nice" for large n.

Figure 8.23

It suggested evaluating `SeriesCoefficient[Exp[Cos[x]], x]` followed by a discussion of what it expected Mathematica® to return (a rather complicated infinite series), but what actually happened was:

In[·]:= `SeriesCoefficient[Exp[Cos[x]], {x, 0, n}]`
Out[·]:= `SeriesCoefficient[E^Cos[x], {x, 0, n}]`

We have already several times made the observation that some LLMs seem to be able to evaluate the Wolfram Language code they write and others do not. In addition, it seems likely that some of them choose not to do so even if they can, and wait for the user to do it and let them know the result. In our experience, the ability to carry out evaluations is useful, makes LLMs more efficient, and sometimes even helps them to return better answers. The above examples seem to suggest to us that WNA was able to evaluate its code (although it sometimes chose not to do so) while o1 was not able to (in none of the above examples it appeared to have done so). However, to be sure, we asked it directly, and it asserted that it was indeed able to perform its own evaluations. It readily agreed that it was a good idea to actually evaluate the code before commenting on the answer and offered to demonstrate this on an example. Then it tried to demonstrate that by performing an evaluation, but the result never came out. When asked to try again, it did, but exactly the same thing happened—no output after a very long wait. Thus we have not been able to resolve this issue, and it remains a puzzle.

To conclude, we are convinced that OpenAI o1 with Mathematica® can be an excellent tool for students and also for instructors. Students, of course, have to treat it with caution and beware of the problems we tried to describe above and possibly others that we have not encountered. One of the many potential uses is quickly creating graphic illustrations in Mathematica® of various mathematical concepts. Another use is a source of exercises to be solved. It appears to be particularly good at creating problems similar to or of the same type as a given problem. This can be, of course, of great help to educators, particularly because LLMs are generally good at avoiding problems, solutions to which can be found online. Of course, the LLMs' ability to solve such problems is going to be itself a problem, especially in relation to homework and exams, but this is an issue that extends far beyond the field of mathematics and is beyond the scope of this book.

We end with an example involving a problem from Chapter 6, concerning the convergence of a sequence of functions, which illustrates some of the issues discussed above.

Consider the sequence of functions f[n] defined on the set of all real numbers given by the formula f[n][x] = x ArcTan[n x], where n runs over the positive integers. Is this sequence uniformly convergent? Can you make an illustration in Mathematica showing the convergence (using Manipulate) so that changing the index n we can see how the functions f[n] approach the limit over some part of the real line? Can you suggest a couple of other problems of this kind (not entirely trivial)?

Below is a step-by-step discussion of the convergence of

$$f_n(x) = x\,\text{ArcTan}(n\,x)$$

as $n \to \infty$ (for $x \in \mathbb{R}$), plus an illustrative Manipulate demonstration in the Wolfram Language. Finally, you'll find two additional (nontrivial) example sequences whose convergence behavior you might want to analyze similarly.

---

1. Analyzing the Limit Function and Uniform Convergence

1. Pointwise Limit

For each fixed real x, consider the behavior of ArcTan(n x) as n grows:
- If $x > 0$, then $n\,x \to +\infty$, so $\text{ArcTan}(n\,x) \to \pi/2$;
  thus $f_n(x) \to x \times (\pi/2) = \pi x/2$.
- If $x = 0$, then $f_n(0) = 0$ for all n, so the limit is 0.
- If $x < 0$, then $n\,x \to -\infty$, so $\text{ArcTan}(n\,x) \to -\pi/2$;
  thus $f_n(x) \to x\,(-\pi/2) = -\pi x/2 = (\pi/2)|x|$ (since x < 0).

An equivalent single expression for the limit is
$f_\infty(x) = (\pi/2)\,|x|$.

2. Uniform Convergence

To check uniform convergence on $\mathbb{R}$, we must show that

$$\sup_x |f_n(x) - f_\infty(x)| \to 0 \text{ as } n \to \infty.$$

A handy bound comes from the asymptotic expansion of ArcTan(u) for large u:

$$\text{ArcTan}(u) = \pi/2 - 1/u + O(1/u^3) \text{ as } u \to +\infty.$$

For $x > 0$ (similarly for $x < 0$ by symmetry), we get
$$f_n(x) - f_\infty(x)$$
$$= \qquad x\,\text{ArcTan}(n\,x) - (\pi/2)\,x$$
$$= x \qquad \text{ArcTan}(n\,x) - \pi/2 \qquad .$$

But for large n x,

$$0 \le \pi/2 - \text{ArcTan}(n\,x) \approx 1/(n\,x),$$

so

$$x\,|\text{ArcTan}(n\,x) - \pi/2| \le x\,(1/(n\,x)) = 1/n,$$

independently of x (as soon as n x is sufficiently large, and also one can check small x directly). Hence

$$\sup_x |f_n(x) - f_\infty(x)| \le 1/n,$$

and $1/n \to 0$. Therefore, f_n converges uniformly to $f_\infty(x) = (\pi/2)|x|$ on $\mathbb{R}$.

---

2. Illustration with Manipulate

Here is a quick example of how you might visualize $f_n(x)$ and its limit in Mathematica. This will produce a slider for n and plot both $f_n(x)$ and the limit function:

```
Manipulate[
  Plot[
    {x ArcTan[n x], (Pi / 2) Abs[x]},
    {x, -2, 2},
    PlotRange -> {-3, 3},
    PlotLabels -> {Automatic, Automatic},
    PlotLegends -> {"f[n] (x) = x ArcTan(n x)", "Limit = (π/2) |x|"}
  ],
  {n, 1, 50, 1}
]
```

⊙ Insert and evaluate   💡 Insert   📋 Copy

- Move the slider for n. As n grows, the plot of x ArcTan(n x) more and more closely "hugs" the V-shaped curve $\pi/2\,|x|$ (though it's slightly below it for x ≠ 0, converging uniformly).
- You can adjust the plot range or the maximum n to taste.

---

3. A Couple of Other Nontrivial Examples

Below are two similar "sequences of functions" problems that go slightly beyond trivial cases:

1. **Example:**
   $g_n(x) = x^2\,e^{\wedge}(-nx)$ for $x \ge 0$; and $g_n(x) = 0$ for $x < 0$.
   - Investigate pointwise limits as x varies. Does it converge to 0 for all x ≥ 0, or are there exceptions?
   - Is the convergence uniform on [0,∞)? On $\mathbb{R}$?
2. **Example:**
   $h_n(x) = (n\,x)/(1 + n^2 x^2)$.
   - Check pointwise behavior. For x ≠ 0, as n → ∞, you might see interesting transitions.
   - Discuss uniformity on $\mathbb{R}$. You can again attempt bounding arguments or find maximum values explicitly.
   These examples illustrate how sometimes the behavior around x=0 or as x→∞ can determine whether convergence is uniform. They also provide practice in bounding or comparing function expressions to show supnorm convergence (or failure thereof).

Summary: The sequence $f_n(x) = x\,\text{ArcTan}(nx)$ converges (pointwise) to $f_\infty(x) = (\pi/2)\,|x|$, and does so uniformly on the entire real line. The Manipulate snippet above can help visualize how quickly (and smoothly) the sequence approaches that limit for all x.

Figure 8.24

```
In[ ]:= Manipulate[
       Plot[
         {x ArcTan[n x], (Pi / 2) Abs[x]},
         {x, -2, 2},
         PlotRange → {-3, 3},
         PlotLabels → {Automatic, Automatic},
         PlotLegends → {"f[n] (x) = x ArcTan(n x)", "Limit = (π/2) |x|"}
       ],
       {n, 1, 50, 1}
     ]
```

Figure 8.25

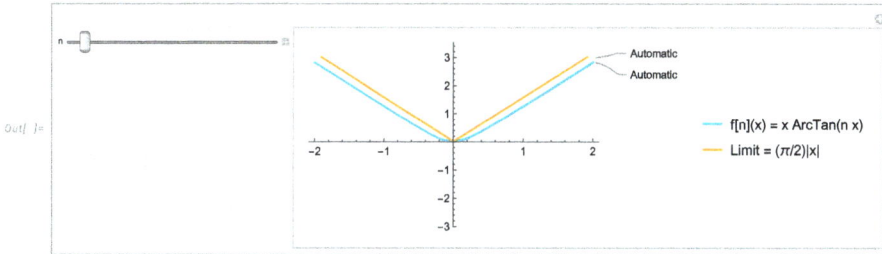

Figure 8.26

Bibliography

[1]   Apostol T. *Calculus. Volume 1: One-variable calculus, with an introduction to linear algebra*, 2nd edition, John Wiley & Sons, Inc. New York, 1967.

[2]   Balser W. *From divergent power series to analytic functions. Theory and application of multisummable power series*. Lecture Notes in Mathematics, 1582, Springer-Verlag, Berlin, 1994.

[3]   Boas RP, Boas HP. *A primer of real functions*, 4th edition. Mathematical Association of America Textbooks, 13, The Carus Mathematical Monographs, The Mathematical Association of America, 1996.

[4]   Bridger M. *Real analysis. A constructive approach*. Pure and Applied Mathematics, New York [John Wiley & Sons], Wiley-Interscience, Hoboken, NJ, 2007.

[5]   Bronstein M. Symbolic integration tutorial. Course notes of an ISSAC'98 tutorial. Available at www-sop.inria.fr/cafe/Manuel.Bronstein/publications.

[6]   Bronstein M. *Symbolic integration I. Transcendental functions, algorithms and computation in mathematics, Vol. 1*, Springer Verlag GmbH, Berlin Heidelberg, 1997.

[7]   Dieudonne J. *Foundations of modern analysis. Enlarged and corrected printing*, Academic Press, New York and London, 1961.

[8]   Geddes KO, Czapor SR, Labahn G. *Algorithms for computer algebra*. Kluwer Academic Publishers, Boston/Dordrecht/London, 1992.

[9]   Grozin A. *Introduction to Mathematica$^{®}$ for physicists*. Graduate Texts in Physics, Springer International Publishing, 2014.

[10] Hardy GH. *Divergent series*. Clarendon Press, Oxford, 1949.

[11] Körner TW. *A companion to analysis. A second first and first second course in analysis*. Graduate Studies in Mathematics, 62, American Mathematical Society, Providence, RI, 2004.

[12] Lang S. *Undergraduate analysis*. Undergraduate Texts in Mathematics, Springer-Verlag, New York, 1997.

[13] Moszyński M. Analiza matematyczna dla informatyków. Wykłady dla pierwszego roku informatyki na Wydziale Matematyki, Informatyki i Mechaniki Uniwersytetu Warszawskiego, 2010 (in Polish). Available at https://www.mimuw.edu.pl/~mmoszyns/Analiza-dla-Informatykow-2017-18/SKRYPT.

[14] Ponnusamy S. *Foundations of mathematical analysis*. Birkhäuser/Springer, New York, 2012.

[15] Suppes P. *Axiomatic set theory*, 1st edition. Dover Books on Mathematics, Dover Publications, 1972.

[16] Wagner DB. Dynamic programming. The Mathematica Journal 5, 42–51, 1995. https://www.mathematica-journal.com/issue/v5i4/columns/wagner/42-51wagner.mj.pdf.

[17] Wagon S. *Mathematica$^{®}$ in action: problem solving through visualization and computation*, 3rd edition, Springer, 2010.

https://doi.org/10.1515/9783111533063-009

Index

** 15
$Assumptions 7, 8
$IterationLimit 57
$MachinePrecision 61
$MaxPrecision 67
$MinPrecision 67
$RecursionLimit 58

Abel's continuity/limit theorem 177
Abel's summation formula 91
Abel's test 92
Abs 147
absolute convergence 80
Accumulate 74, 75
Accuracy 18
almost uniform convergence 202
Analytic 143
antiderivative 215
arbitrary precision arithmetic 61
ArcSinh 161
ArcTan 189, 223
Arg 223
Assuming 7, 9
Assumptions 7–9, 137, 232, 233, 238
Asymptotic 105, 187
AsymptoticLess 82
AsymptoticLessEqual 82, 83
Attribute 146

Block 25, 27, 57, 58, 68, 146, 232
BlockRandom 228

Cauchy principal value of the integral 239
Cauchy product 101
Cauchy–Hadamard Theorem 181
Cauchy's root test 81
Cesàro regularization 102
ChatObject 250
Clear 25
ColorFunction 173
ColorFunctionScaling 173
comparison test 81
ComplexExpand 12, 31, 48, 223
ComplexInfinity 16
ComplexityFunction 4, 6, 89
ComplexPlot3D 197
concave 170
Constant 146

Constants 145
convex 170
Cosh 159
critical point 152
critical value 152
CubeRoot 142

D 145, 148
D'Alembert's ratio test 81
Darboux integrable 227
Darboux integral 227
Defer 20
derivation 220
derivative 138
Derivative 147, 148
difference quotient 136
DifferenceQuotient 137
Differences 68, 74, 75
differentiable 138
differential equation 215
differential field 220
differential operator 220
DirectedInfinity[ ] 16
DirectedInfinity 16
Direction 95, 111
Dirichlet's test 92
Dirichlet's test for uniform convergence 209
discontinuous functions 117
discrete dynamical system 63
DiscreteLimit 28, 38, 77, 109
DiscreteMaxLimit 39
DiscreteMinLimit 39
DiscretePlot 121
DiscretePlot3D 48
DiscreteRatio 89
Distribute 14
domains 2
Dot 90
DSolve 215
Dt 145
dynamic programming 55

Element 2, 34
Exists 10
Expand 3, 14
extension of a differential field 220
extremum 152

https://doi.org/10.1515/9783111533063-010

Factor 5
FactorSquareFree 223
Fermat's Interior Extremum Theorem 152
Fibonacci[n] 65
field of constants 220
FindRoot 37, 71, 124
fixed point 63
FixedPoint 60
FixedPointList 60
FoldList 75
ForAll 10
FromDigits 18
FullSimplify 3, 10
FunctionConvexity 174
FunctionExpand 5, 44, 262
FunctionMonotonicity 173
FunctionRange 160
Fundamental Theorem of Calculus 231

GenerateConditions 110
GeneratedParameters 215
greatest lower bound 45

HarmonicNumber 44
Head 184
HoldForm 19

If 118
ImplicitD 168
improper integrals 237
Inactivate 19, 88
Inactive 216
indefinite integral 215
Indeterminate 16
induction axiom 21
inexact number 60
infimum 45
Infinity 16
InputForm 60
IntegerDigits 18
IntegerString 18
Integral test 82
integral test 241
IntegralTest 242
Integrate 215, 230, 233, 235
IntegrateChangeVariables 234
Intermediate value property 122
Interval 18
InverseFunction 128, 129, 159, 187
InverseSeries 187

Jensen's inequality 176
Join 1

Lagrange Remainder Terms Theorem 190
LeafCount 4
least upper bound 45
left differentiable 139
left hand derivative 139
Leibniz's test 93
Limit 28, 38, 77, 109, 110, 126, 213
limit comparison test 82
limit point 109
Liouville's Principle 220
Lipschitz condition 132
Lipschitz function 152
Listable 22
local maximum 151
local minimum 151
lower bound 45
lower Darboux sum 227
LucasL 66

MachinePrecision 60, 61, 67, 103, 199
Maclaurin series 183
Manipulate 64, 164
Maximize 46, 47, 49, 154
MaxIterations 72
MaxLimit 39, 181
MaxRecursion 72
MaxValue 125
Mean 156
Mean Value Theorem 152
MemberQ 2
Method 81, 82, 96, 242
Minimize 46, 47, 49, 154
MinLimit 39
MinRecursion 72, 73
Module 25, 27, 146

N 71
N[a, p] 61, 67
Nest 56, 69
NestList 56
Newton–Leibniz formula 231
NIntegrate 72, 233–235, 240
NMaximize 51, 52
norm 201
Normal 184, 224
NSolve 31, 124

NSum 78–80
NSumTerms 80
null sequence 74
NumberForm 240

O 183

PadeApproximant 107
Partition 156
partition 226
Peano Remainder Terms Theorem 182
Peano's axioms 20
Piecewise 117, 118, 120, 216
Plot 114, 173
Plot3D 48
pointwise convergence 202
PolyLog 178, 225
PolynomialQuotentRemainder 222
Power 142
power series 103, 177
PowerExpand 9
Precision 18, 60
Prime 22, 28
PrimePi 24, 29
PrimeQ 24
primitive function 215
PrincipalValue 239
Principle of Mathematical Induction 21
ProductLog 164
ProvablePrimeQ 24

Quiet 64

Raabe's test 82
Rationalize 191
Reap 41
RecurrenceTable 55
recursive sequence 55
Reduce 10, 30, 84, 155
Refine 7
RegionPlot 229
Regularization 101
regularized sum 101
Remove 25
Resolve 10
Return 254
Riemann integrable 229
Riemann integral 229
Riemann sum 235

Riemann's theorem 98
right differentiable 139
right hand derivative 139
Risch algorithm 217
Root 30, 32
RootReduce 32, 34
Roots 31
RootSum 224
RSolve 57, 62, 131

SeedRandom 227
Select 41
sequence 37
series 76
Series 107, 177, 183, 184
SeriesCoefficient 106, 184, 185
SeriesData 184
sets 1
Sign 68, 156
Simplify 3, 40
Sinh 159
Solve 30
Sort 1, 41
Sow 41
StrictInequalities 174
Sum 76–78, 101
sum of the series 76
SumConvergence 78, 81, 82, 95, 104, 179, 188, 242
supremum 45
Surd 141

Table 40
TargetFunctions 13
Taylor polynomial 181
Taylor series 181
the least upper bound 45
The Weierstrass theorem 122
Thread 41
TimeConstraint 3
Timing 59
ToRadicals 32
ToRules 31
Trace 26
TraditionalForm 39, 142
TransformationFunctions 4, 5
TreeForm 4
TrigReduce 4
TrigToExp 220, 222
TruncateSum 77

uniform convergence 202
uniformly continuous 131
Union 1, 2
upper bound 45
upper Darboux sum 227

VerifyConvergence 78

Weierstrass M-test 209
While 123, 191
With 25, 27
WorkingPrecision 80, 113